Beyond the Big Ditch

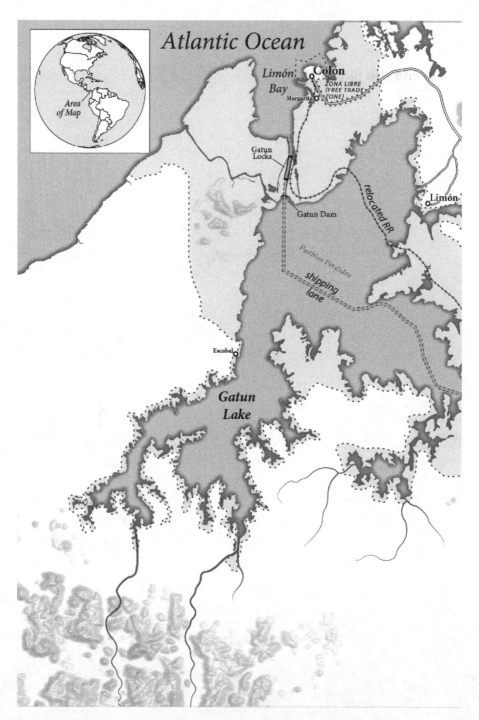

Figure 0.1
Credit: Tim Stallman, used with permission.

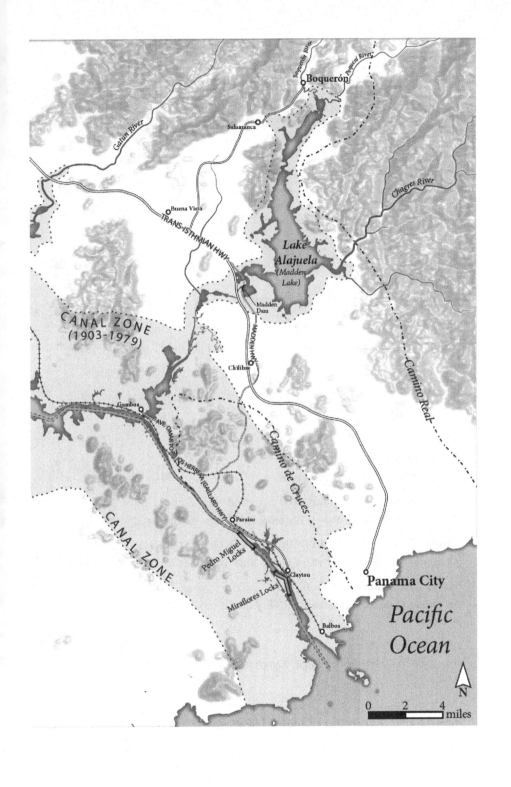

Infrastructures Series
edited by Geoffrey Bowker and Paul N. Edwards

Beyond the Big Ditch

Politics, Ecology, and Infrastructure at the Panama Canal

Ashley Carse

The MIT Press
Cambridge, Massachusetts
London, England

This book was set in Stone Sans and Stone Serif by Toppan Best-set Premedia Limited.

Library of Congress Cataloging-in-Publication Data

Carse, Ashley, 1977–

Beyond the big ditch : politics, ecology, and infrastructure at the Panama Canal / Ashley Carse.

pages   cm. – (Infrastructures series)

Includes bibliographical references and index.

ISBN 978-0-262-02811-0 (hardcover : alk. paper), 978-0-262-53741-4 (pb)

1. Panama Canal (Panama)–Design and construction.   2. Panama Canal (Panama)–Social conditions.   3. Panama Canal (Panama)–Environmental aspects.   I. Title.

TC774.C4365   2014

972.87'505–dc23

2014008289

For Sara

Contents

Preface

I arrived in Panama City in 2006 with an invitation to collaborate on an interdisciplinary environmental research project at the Smithsonian Tropical Research Institute. At that time, I was a graduate student in anthropology considering a dissertation project in Panama. I knew little about the Panama Canal and considered it too "big" for anthropological research. Anthropology, after all, has pursued some of the largest human questions in the smallest places: islands, villages, and neighborhoods. In this tradition, I planned to study environmental politics in two rural communities near the Panama Canal. It seems obvious now, but it took me some time to realize that I could not understand politics and ecology around a massive engineering project while ignoring infrastructure.

When I began conducting interviews in rural communities, people often redirected my questions about the environment—focused on trees, cattle, and agriculture—to discuss the condition of gravel roads, power lines, and potable water systems. I wanted to understand their experiences with environmental policies, but they were more interested in discussing the politics of the built environment. Where does potable water (not) run? Where are roads (not) built and maintained? Where does electricity (not) work? The answers to these questions, I learned, are inextricably bound up with the story of transportation across Panama.

Global trade depends on cheap and efficient transportation, which, in turn, requires networks of canals, roads, airports, pipelines, railroads, dredged rivers, harbors, ports, and telecommunications facilities. On the one hand, these infrastructures are means of overcoming constraints to movement and exchange presented by the physical environment (distance, topography, hydrology, and climate). On the other hand, the construction, management, maintenance, and expansion of transportation infrastructure demands extraordinary amounts of land, fresh water, and natural resources. Places—entire regions—are remade in the image of transportation in ways that are often invisible to the millions, even billions, of producers and consumers around the world that depend on them.

The anthropologist Clifford Geertz writes, "As a chameleon tunes himself to his setting, growing into it as though he were part of it, just another dun rock or green leaf, a society tunes itself to its landscape ... until it seems to an outside observer that it could not possibly be anywhere else than where it is, and that, located where it is, it could not be otherwise than what it is."[1] My fieldwork in rural communities at the margins of the Panama Canal taught me to see the trade route as part of the regional landscape and everyday life, as opposed to a "big ditch" or technical system——a perspective that turned out to be revealing.

Through a historical and ethnographic analysis of the changing relationships that bind the Panama Canal and its rural hinterlands, this book shows how infrastructures both transform and depend on the ecologies that they cross. Thus, the reorganization of landscapes to facilitate global flow also produces friction: new canals and reservoirs flood old hunting trails and river routes, new railroad tracks bypass the economies of river towns, and new highways fracture urban neighborhoods. Until we recognize that environmental conflicts are shaped by how our infrastructures are embedded in landscapes and bound up with long networks, we will have a difficult time developing solutions that fit with how the communities they connect experience the world and what they consider fair.

Acknowledgments

I have accrued many debts since I began this project. I want to start by thanking those in Panama who welcomed me and took the time to speak with me, particularly the families that opened their homes and gamely tolerated my questions. I also owe a debt of gratitude to two Panamanian scholars who shaped this work. I want to thank my friend Francisco Herrera, an anthropologist at the University of Panama, for being a generous ally, an amazing reference on all things Panamanian, and for commenting on an earlier manuscript of the book. I also want to thank Stanley Heckadon-Moreno at the Smithsonian Tropical Research Institute (STRI), who has done more than anyone to explain the complex social and environmental transformations of rural Panama over the last century. This work would not have been possible without his foundational research and the hours we spent talking in his office or over coffee at the Tupper cafeteria. STRI provided a welcoming institutional home on the isthmus. I thank everyone there, especially Jeff Hall, Eva Garen, Vielka Chang-Yau and the library staff, and the late Neal Smith, for their support over the years.

I owe a great deal to the mentors who oversaw the development of this project. Peter Redfield has been a model of rigor, creativity, and professionalism. Since graduate school, he has fostered my intellectual and professional development with pragmatism, generosity, and uncommon wit. Carole Crumley has exemplified interdisciplinary engagement in her efforts to bridge C. P. Snow's "Two Cultures." She taught me to see landscape as history and history as landscape. Arturo Escobar deepened my understanding of the politics of theory, teaching me to pay attention to the possibilities opened by thinking differently. Wendy Wolford introduced me to political ecology and reminded me that production matters. Julie Velásquez Runk provided important contacts on the isthmus, key references, and a thoughtful reading of my dissertation. Her tireless commitment to engaged scholarship in Panama has been an inspiration.

I would also like to thank the many friends and colleagues who read and commented on writing that became parts of the book: Samer Alatout, Nikhil Anand, Hannah Appel, Karen Bakker, Andrea Ballestero, Jessica Barnes, Wiebe Bijker, Stephanie Friede, Liz Hennessy, Josh Lewis, Mike Lynch, Alex Nading, Mihir Pandya, Dana Powell, Janell Rothenberg, and Austin Zeiderman. My thanks go as well to a cohort of fellow travelers in Panama: Kurt Dillon, Christine Keiner, Eben Kirksey, Jeff Parker, Megan Raby, Matt Scalena, Blake Scott, Ezer Vierba, and Katie Zien. I thank Blake, in particular, for his generous comments and insights on the manuscript. I am also grateful to my friend Tim Baird, who has challenged me to clearly articulate my thoughts about people and the environment for nearly a decade now.

The ideas in the book developed through presentations and workshops at a number of institutions. I'd like to thank the participants in the 2010 Workshop for the History of Environment, Agriculture, Technology, and Science at the University of Wisconsin and the "From Field to Table" conference at the University of South Carolina for their comments on an earlier version of chapter 8. I am also grateful to all of the participants in the "Infrastructural Worlds 2014" workshop, sponsored by the Duke University Department of Cultural Anthropology, for two days of generative conversations on infrastructure. In particular, Majed Akhter, Stephanie Kane, Ben Mendelsohn, and Chitra Venkataramani provided useful feedback on the book's introductory chapter. Audiences at colloquia sponsored by the Environmental Studies Program at Trinity University, Department of Anthropology at the University of Denver, Department of Anthropology at the University of California, Santa Cruz, and Center for Science, Technology, and Society at Drexel University also raised useful questions and critiques.

This book would not exist without the institutions that funded my research and writing. A NSF-IGERT Fellowship in Population and Environment at the Carolina Population Center at the University of North Carolina supported the first three years of my doctoral education and provided me with a foundational interdisciplinary training. A UNC Latin American Studies Tinker Research Grant supported preliminary research in Panama. The Fulbright Program, Wenner-Gren Foundation (Grant #7701), and a UNC Mellon-Gil Dissertation Fellowship supported my dissertation research. Archival research in the United States was funded by the Wenner-Gren Foundation and a Smithsonian Short-Term Research Fellowship. While at the Smithsonian Institution Archives in Washington, D.C., I was fortunate to be hosted by Pamela Henson and to interact with Jeffrey Stine. Early writing support was provided by a UNC Dissertation Completion Fellowship. I completed the manuscript thanks to a National Science Foundation Postdoctoral Fellowship in Science, Technology, and Society (Award #1257333). Any opinions, findings, and conclusions or recommendations expressed in this material are my own and do not necessarily reflect the views of the National Science Foundation or my other funders.

While writing and revising this text, I benefited from time at institutions on both coasts. I spent two years happily teaching and writing near the Pacific. At Whittier College, I'd like to thank my mentors sal johnston and Ann Kakaliouras and my students, who taught me about the political ecology of California. I'd also like to thank my colleagues Julie Collins-Dogrul, Susan Gotsch, David Iyam, Bob Marks, Becky Overmyer-Velazquez, and Nat Zappia for making my time at Whittier so pleasant. While in Los Angeles, I also benefited from participation in the Infrastructure and Environment Working Group convened through the Science, Technology, and Society Program at the University of Southern California. Finally, I'd like to thank my new colleagues at the University of Virginia, particularly Deborah Johnson, Ed Berger, and Fred Damon, for all of their support.

At the MIT Press, I would like to thank Margy Avery for her enthusiasm about the project; Paul Edwards and Geoffrey Bowker, editors of the Infrastructures series, for providing smart and timely guidance; and Marcy Ross, my excellent manuscript editor, for her useful suggestions and patience. I owe a huge debt to all four anonymous manuscript reviewers for their extremely generous and thoughtful comments.

Elements of chapters 1, 3, and 6 appeared in a different version in Ashley Carse, "Nature as Infrastructure: Making and Managing the Panama Canal Watershed," *Social Studies of Science* 42, no. 4 (2012): 539–563.

I could not have completed the book without the love and support of my parents, Jim and Lib, and my brothers, Nick and Steve. Finally, and most importantly, my gratitude goes to Sara for her love, wisdom, patience, and endurance. She has lived with the places, people, and ideas that appear in this book for a long time and read every word that follows multiple times—never failing to ask, "But what's at stake here?" And thanks to our son, Eli, for reminding me just how much is at stake and how vibrant life can be.

1 Introduction: The Machete and the Freighter

Figure 1.1
Southbound ship passes through the Panama Canal's Miraflores Locks. Photo by
Tim Baird. Used with permission.

If you travel to Panama and want to see its canal, you will likely end up
at the Miraflores Locks Visitor Center on the outskirts of Panama City. The
building's interior—all cream adobe walls, towering plate glass windows,
and marble floors—is a showcase for the history of the iconic trade route.
It contains a museum, a theater, and a gift shop selling canal neckties and

tea sets. But the real action is outside. From a three-tiered viewing deck, tourists watch a parade of container ships, tankers, and cruise ships pass through locks built a century ago.

In 2008, I visited the Miraflores Locks on Earth Day. The announcer narrated the action from the viewing deck, his voice blaring from speakers overhead. Reading in English and Spanish from a computer printout, he explained how the canal works and listed facts about the colossal size, enormous cost, and impressive durability of its aging technical system—the locks, dams, and gates that correspond with the popular under-standing of infrastructure as hardware. When he mentioned the Earth Day celebration, however, he introduced us to a different canal—a sprawling, living organism. Moving beyond the engineering achievements of the early twentieth century, he turned to the critical, ongoing, often over-looked environmental dimensions of transportation. Pointing to the source of canal water in the mountainous headwaters of the Chagres River, he said, "We can't underestimate the importance of protecting the envi-ronment, ladies and gentlemen. Here it's basic and simple: no rainforest, no canal. That's how we replenish the great amount of water we use to move vessels."

The Panama Canal is a fifty-mile long waterway organized around an aquatic staircase of six locks—three steps up, three down—that requires an enormous volume of fresh water to function (figures 1.2 and 1.3). The locks are located at Miraflores (two locks, Pacific side), Pedro Miguel (one lock, Pacific side) and Gatun (three locks, Atlantic side). Here, crudely, is how they work: an Atlantic-bound ship slides into the lower chamber at Miraflores and massive steel gates swing slowly closed behind it. Then, a lockmaster opens valves upstream—the next "step" up the staircase—and gravity pushes water through culverts the size of subway tunnels embedded in the concrete chamber walls. It flows into cross-culverts beneath the chamber floor and then erupts upward. The water in the chamber rises, lifting the ship with it. This lockage process is repeated in a second chamber at Miraflores and then at Pedro Miguel, the next lock complex, before the ship travels through the Culebra Cut and across Gatun Lake at eighty-five feet above sea level. At the Gatun Locks, near Colón, the ship is lowered back to sea level and enters the Atlantic. The order of transit is reversed for Pacific-bound ships.

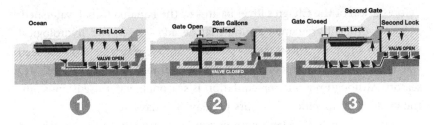

Figure 1.2
How the lock canal system works. Image by David Kuhn, used with permission.

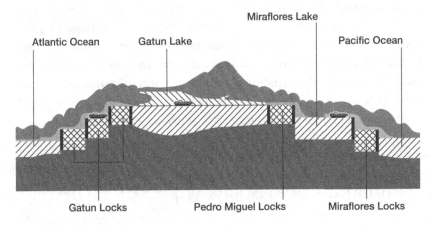

Figure 1.3
Ships traverse an "aquatic staircase" of locks as they transit between the seas. Image by David Kuhn, used with permission.

While passing through the locks, each ship that transits the canal drains an astounding *52 million gallons of fresh water* into the oceans—approximately the same volume of water as 78 Olympic swimming pools, 15 million toilet flushes, or the daily domestic consumption of half a million Panamanians.[1] Where does all of this water come from? Like many lock canals, Panama's interoceanic waterway is supplied with vital liquid by river-fed, summit-level reservoirs—Gatun Lake and Alajuela Lake—that store water for navigation and a number of other uses. Because the locks transform Chagres River water into money and jobs, the supply available for canal use was—and remains—a concern in Panama and beyond.

"No rainforests, no canal," the announcer at Miraflores said. What did he mean? Since the mid-1970s, some scientists have warned that the deforestation of the Chagres River basin—or the Panama Canal watershed, as it was officially renamed during that decade—posed two hydrological threats to shipping: the siltation of water storage reservoirs and reduced water flow into the reservoirs during the annual January-to-April dry season. Although regional precipitation is seasonal, water use is constant: the canal is in operation 24 hours per day, 365 days per year.

Canal administrators became increasingly concerned about watershed deforestation after a long drought in 1977 (figure 1.4). As the water level in the canal dropped, some ships had to reduce their loads to transit a shallower canal or were diverted thousands of miles south around Cape Horn. In 1978, a US forester published an influential article predicting the "death" of the Panama Canal as a reliable world trade route if its engineered infrastructure was not integrated with environmental management.[2] By discursively extending water management infrastructure to rural areas upstream, claims about the hydrological properties of forests initiated the transformation of life and landscape across an entire region.

More Than 62 Years of Service to World Commerce

The Panama Canal Spillway

Vol. XV, No. 26   Friday, February 11, 1977

Shooting a sound-on-film interview with Gov. H. R. Parfitt is a camera crew from Telepress International News, Teheran, Iran, one of an increasing number of United States and foreign reporters and cameramen arriving on the Isthmus to file on-the-spot stories from the Canal Zone. Other newspapers, news magazines and television networks to send representatives to the Canal Zone in recent days included NBC, ABC, Los Angeles Times, Newsweek and the Milwaukee Journal.

Dry Conditions Continue To Set More Records Here

The unusually dry weather that prevailed across the Isthmus during the past rainy season continued through December and January, establishing new lows for water inflows into Gatun Lake for both months.

Accumulated rainfall at the 19 stations across the Isthmus during December was only 1.94 inches, 77 percent below the average December accumulation of 8.26 inches. In January, the rainfall totalled 1.47 inches, 60 percent below the average of 3.69 inches normally recorded during that month.

The water inflow to Gatun Lake during December was 4,140 million cubic feet (mcf), far below the previous record low of 5,781 set in December 1918 and was 82 percent less than the average for that month.

The January inflow of 1,132 mcf shattered the previous January mark of 1,284 that was set in 1975 and amounted to only a fraction of the 63-year January average of 6,980 mcf. The level of inflows into the lake is a reflection of the quantity of rainfall in the basin during the previous month.

Weather experts with the Meteorological and Hydrographic Branch explain that the dry statistics for December and January can be expected to continue now that dry season has settled in. They point out that the reason for the record breaking figures for December and January is that the current dry season arrived a full month earlier than usual. December is normally considered a rainy month but the past December was entered into the dry season column and the January figures for rainfall and inflow resembled those normal for February.

Panama Official Invited To View PC EEO Program

Bruce A. Quinn, Director of Equal Employment Opportunity for the Panama Canal organization, has invited a key Panamanian official to join him and the residents and employees of the Canal Zone for the events scheduled to take place during Black History Month.

In his letter to Pedro Brin Martinez, Quinn wrote: "I fully appreciate your efforts toward the safeguarding of Human Rights in your capacity as representative of the Republic of Panama to the U.N. Committee for the Elimination of All Forms of Discrimination. As Director of Equal Employment Opportunity for the Panama Canal

Figure 1.4
Drought conditions in 1977 led to a shortage of water for navigation. Source: *Panama Canal Spillway*, Vol. XV, No. 16, February 11, 1977.

Reflecting on the rise of watershed management around the canal in the 1980s, the Panamanian sociologist Stanley Heckadon-Moreno wrote: "In essence, the [Panamanian] government decided that, in order to save the canal, the forests had to be protected from the machetes of the farmers."[3] Save the Panama Canal from machetes? How could such a simple "local" technology threaten an icon of the seemingly inexorable march of modern technology and global connection? While US state and military officials had long prepared for a foreign assault on the canal's locks and dams, they had never worried that interoceanic shipping might be disrupted by the everyday agricultural labor of Panamanian campesinos (rural people) making a meager living on land many miles from the shipping lane. Yet, in the 1970s and 1980s, an alarmist discourse scripted campesino land use practices as a serious water management problem and the locus of environmental interventions— namely, conservation policies and agricultural restrictions—that became part of the behind-the-scenes work of interoceanic transportation across Panama.

Here, then, are the paradoxes at the core of this book: (1) *the Panama Canal needs forests* and (2) *the canal had to be saved from machetes*. Beginning with an examination of the conditions of possibility for those two statements, the book examines both technical and governmental efforts to establish landscapes conducive to transportation, foregrounding the lived consequences of those efforts at their rural margins. In so doing, I develop four broader arguments about politics, ecology, and infrastructure.

First, infrastructure is not a specific class of artifact or system, but an ongoing process of relationship building. Seen in this way, engineered canals and highways are surprisingly social and ecological. As temporary lines across active environments that erode, rust, and fracture them, infrastructures advance and retreat in relation to the capital and labor channeled into their construction and maintenance.

Second, infrastructures have grown long enough to encircle—if not encompass—the planet. These global infrastructures channel flows of people, goods, and wealth through certain areas (like the Chagres River basin) with feverish intensity, placing great demands on neighboring populations and environments. However, people and places are not "globalized" in any final way. Over time, they can be connected and disconnected, integrated and bypassed.

Third, infrastructures produce environments, and vice versa. On the one hand, reservoirs, wetlands, reefs, forests, and other "natural" landscapes may be organized in ways that reflect the design, management, and politics of technical systems. On the other hand, the ecologies that accrete around infrastructures are irreducible to environmental effects or services. As landscape, infrastructures give rise to political ecologies with winners and losers.

Fourth, environmental conflicts may emerge at the intersection of competing global infrastructures that organize landscapes to "serve" different purposes, economies, or communities. Around the Chagres River, for example, transportation and rural development infrastructures have come into tension with one another, reconfiguring expectations, responsibilities, and social practices.

Beyond the Big Ditch

The Panama Canal has borne fewer books than ships, but enough have been written that another one demands justification. For generations, North American historians have told the story of a monumental engineering project begun in 1904 and completed in 1914 by virtue of US political will, scientific and technological innovation, and migrant labor. These histories often begin with early colonial fantasies of discovering a hidden strait across Panama linking Europe and Asia by water. Then, they chronicle centuries of efforts by foreign governments and capitalists to construct interoceanic routes—a Spanish colonial road, a North American railroad, a failed French sea-level canal—in a manner that sets the stage for the successful opening of the waterway by the US government. From a certain US or global perspective, then, the opening of the canal symbolized human progress and the capacity of the modern state to liberate society and economy from worldly constraints through engineering.[4]

Meanwhile, a body of critical and revisionist scholarship—established in the Spanish-language historiography and more recent in English—has reframed canal construction and administration as an imperial project built on the backs of black migrant laborers and at the expense of Panamanians. Intellectuals and nationalists in Panama have tended to frame the opening of the canal and enclosure of the US Canal Zone as the beginning of yet another chapter in a saga of colonialism and dependence—as opposed to a universal moment of progress and liberation. They argue that a disproportionate focus on the transport sector (under foreign control since the colonial era) has had a number of negative domestic consequences, including: concentration of wealth, population, and political power in the transit zone, leading to its disintegration with the rest of the country; underdevelopment of the agricultural and industrial sectors; and a dependence on foreign markets that has led to recurring cycles of boom and bust.[5]

What has not been written is a history of the Panama Canal that the Miraflores Locks announcer described: a sprawling, highly demanding organism clothed in treaties, legislation, and institutions. Approaching the canal as infrastructure embedded in the landscape sheds new light on some old narratives.

First, environment: When the Panama Canal opened in 1914, it was touted as modern man's ultimate conquest of nature. Oceanic connection entailed an unprecedented alteration of isthmian geography: moving earth, redirecting a river, and controlling disease. While construction histories are full of torrential rains, floods, mudslides, and disease-carrying mosquitoes, nature seemed to fall out of the picture after excavation was completed. And yet, environmental management by other names— especially maintenance, sanitation, and protection—continued to be central to canal administration. Moving beyond narratives that nature around the canal was either conquered through historical construction or threatened through recent deforestation, I argue that, for over a century, canal administrators have built infrastructure and managed environments in ways that serve specific communities, bringing transportation efforts into tension with other ways of knowing and governing the landscape.

Second, politics: Scholarship and popular writing about the Panama Canal and US Canal Zone tend to focus on the political geography of the nation-state. Whether the focus is international diplomacy, Yankee imperialism, or national sovereignty, many histories of the canal have been framed in terms of a United States versus Republic of Panama binary. I complicate this framing by conceptualizing the state as a material entity constructed through civil engineering efforts, land-use policies, environmental management, and development projects. This emphasis on territorial politics reveals geographies that cross and confound national boundaries, while also reaching out to link to global networks. My focus, then, is how the politics of infrastructures play out across multiple scales as diverse actors have shaped, negotiated around, and adapted to them.

Third: everyday life. For many people living at its margins, the Panama Canal is not experienced as a regional technical system or part of global infrastructure. Instead, it has become, as Chandra Mukerji writes about another engineered waterway, "a brute fact in the countryside ... something to work with and work around like a mountain, not something to debate or query about its history."[6] By reworking the built environment, canal construction and management has both closed down local possibilities and opened up new opportunities. Thus, many rural people know the canal and make sense of their changing relationships with its imagined national and global communities through landscape change—rivers that become lakes, roads built and abandoned, buildings constructed and in ruin, forests that become fields, and fields grown up thick with weeds.

Global Theory in Panama

If anything is global, the Panama Canal must be. Over one million vessels have passed through the waterway since it opened for business in 1914. During the first decade, one to five thousand ships transited per year. In recent years, twelve to thirteen thousand ships have transited annually, carrying about 5 percent of all global trade. This is not to say that the canal links the entire globe. Rather, the waterway primarily serves specific maritime routes (figure 1.5). Nearly 40 percent of canal traffic by tonnage in 2012 traveled between Asia (mainly China) and the eastern United States. Ships plying the second most traveled route, between the eastern United States and western South America, made up 9 percent of canal traffic.[7]

Interoceanic transit is a lucrative business. The Panama Canal Authority, the quasi-autonomous Panamanian state institution that administers the waterway, charges tolls that are calculated according to ship capacity; some fees have exceeded three hundred thousand dollars per transit. The Authority collected 2.4 billion dollars in revenue in 2012, and contributed 1 billion dollars in profits to the Panamanian government, while employing more than ten thousand people. Therefore, regional water shortage has national and international implications.[8]

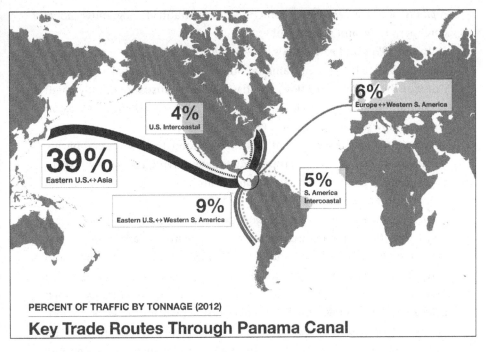

PERCENT OF TRAFFIC BY TONNAGE (2012)
Key Trade Routes Through Panama Canal

Figure 1.5
Image by David Kuhn, used with permission.

How should we conceptualize "the global" in relation to the Panama Canal? Global theories draw attention to particular spatial forms, patterns, and linkages that might otherwise remain unnoticed, but by bringing certain relationships to the foreground they obscure other parts of the landscape.[9] Two key metaphors—structure and flow—have dominated global theory in the social sciences and humanities. In the late 1960s and 1970s, Marxian world-system and dependency theorists conceptualized the world as a totality of interconnected processes or structures knowable through the analytical tools of political economy. These theories were a corrective to a number of prevailing assumptions in the social sciences, not least of which was the tendency of academic disciplines to identify a central concept (society, culture, the state) and then construct a corresponding spatial unit to study. To think, instead, in terms of a single world system was to question the heuristic utility of such boundaries for understanding social organization and to focus instead on political-economic structure. The key argument is that, since 1900, there has been only one *capitalist world system* that is structured by a stratified geography: global core, periphery, and semi-periphery.[10]

In the 1990s and early 2000s, social theorists Arjun Appadurai, Manuel Castells, and Ulf Hannerz, among others, reframed global connection through the metaphor of flow, emphasizing the transnational movement of capital, images, commodities, and people. To think about the global in terms of flow was to emphasize emerging connections—especially across communication networks—that were understood to link geographically distant sites more tightly than proximate ones, giving rise to new experiences of space and time and, significantly, undermining the boundaries of nation-states and places.[11]

Despite critical philosophical and political differences, world-system and global flow theories are often embedded with the assumptions that anthropologists Akhil Gupta and James Ferguson call encompassment and verticality.[12] Encompassment is the notion that scales are nested like a Russian *matryoshka* doll: the individual is situated within a locality, which is located within a region, within a nation-state, within the globe. World-system theory, for example, posits a single planetary system with no outside. Verticality is the notion that scales are organized in a stratified manner, with the nation-state "above" a region or locality and "below" the globe. Power is understood to operate in a top-down fashion. Although the metaphor of flow suggests connection over unity, it remains abstract and reproduces a global-local dualism, in which connection erodes place.[13]

If we begin with these assumptions, then we are likely to ask questions that reproduce their logics, like: How do global forces affect a local community? Or, the same question reversed: How do people resist or refashion global forces? Yet both questions situate agency at "larger" or "higher" scales while treating "smaller" or "lower"—the local, everyday—as either passive or reactive.

With that in mind, let's return to the problem of the machete and the freighter. The relationship between Panama Canal administrators and campesinos was, of course, asymmetrical in terms of capital and political clout, but a David and Goliath narrative does not fit the situation, for the freighter (our global icon) was said to be threatened by the machete (our local symbol), and vice versa. Their relationship was neither a case of global forces eroding local community, nor local resistance, but the tension between two infrastructures that crossed scales.

I suggest that to understand the relationship between the machete and the freighter, we must conceptualize scale as an ongoing achievement rather than an encompassing sphere or hierarchical system—simultaneously more concrete (ships, ports, containers) and more unfinished than it is often imagined to be. Theorists have developed frameworks like technological zones, networked infrastructures, translocality, global assemblages, and friction to foreground how global connection is made through networks or assemblages of technologies, governmental and administrative bodies, value regimes, and cultural forms.[14] I extend this literature by emphasizing the ongoing, place-based environmental management work involved in transportation and showing how multiple "globals"—conceptualized as infrastructures—may colonize the same landscape.

Global Infrastructures Emerge, Connect, and Disconnect

What is infrastructure? The word infrastructure came to English from French in the 1920s, where it referred to substrate material below railroad tracks. We often conceptualize infrastructures as "hard" technical artifacts or systems, rather than processes. However, in its earliest documented English use in 1927, the Oxford English Dictionary defined the new word in active terms, referring to "the tunnels, bridges, culverts, and 'infrastructure' work generally" of French railroads.[15]

I emphasize infrastructural work—a term used by science, technology, and society scholar Geoffrey Bowker[16]—to foreground the variety of organizational techniques (technical, governmental, administrative, environmental) that create the conditions of possibility for rapid and cheap communication and exchange across distance. As scholars have shown, infrastructure does not refer to any specific class of artifact, but to a process of relationship building and maintenance.[17]

Global infrastructures emerge. Unlike regional or national sociotechnical systems constructed by identifiable builders, they grow through the development of *gateways*: standards or technical devices that bridge systems with unique histories and resolve technical, political, and legal issues so that they can be used as a single infrastructure.[18] Geographer Andrew Barry uses the term *infrastructural zones* to describe the emergent and striated spatial forms created by the reduction of differences between systems via common connection standards.[19] The shipping container is an iconic example of a gateway that also played an important role in the history of the Panama Canal.

The economist Marc Levinson argues that the standard shipping container, invented in the mid-1950s, gave rise to contemporary economic globalization by dramatically reducing shipping costs, thus making it profitable to trade across unprecedented distances. The container, an ideal gateway, reduced friction (in terms of time and cost) during transfers across formerly discrete transportation systems with different designs as well as land- and water-based modes of transit.[20] Before the 1960s, when container use proliferated, most tractor-trailers, trains, and ships were loaded and unloaded by hand (except bulk commodities), which demanded large temporary labor forces at ports and other transfer points.

The container was designed to bridge existing transportation systems, but then it began to remake them in its image, initiating a cascade of

Figure 1.6
The world the shipping container made, as seen at the Port of Long Beach. Photo by the author.

technical, economic, and sociopolitical changes. By reducing costs, the container enabled lower shipping rates, which attracted more freight and enabled investments in bigger ships designed to maximize container storage, unlike the preceding generation of repurposed freighters. To attract more business, port administrators modernized and expanded their facilities, deepened their harbors for large container ships, and bigger cranes to reduce handling times for ships.[21]

Infrastructures, as anthropologists have shown, operate at material and poetic registers.[22] If ships and railroads move coal and cars from one place to another, then they are also vessels for cultural imaginaries, desires, promises, and fantasies. As such, infrastructure projects are fertile ground for what anthropologist Anna Tsing calls global conjuring. For her, globalism—a commitment to and public evocation of the global scale—strengthens other scale-making projects.[23] Indeed, port authorities, transportation firms, and governments ask publics to imagine the local port or regional economy in terms of a future delivered by bigger ships or commodity crops—if, and only if, enough capital and politics can be mobilized to capture them, or so boosters claim. In this way, appeals to the global are used to build and remake localities. From this perspective, the globe begins to look like a multitude of scale-making projects reaching outward to attach themselves to different infrastructures, rather than a solid sphere. This was certainly the case in twentieth-century Panama.

The story of the machete and the freighter was shaped not only by long transportation networks, but also by a second, latent global infrastructure: international economic development. The English use of the word "infrastructure" proliferated through the discourse of that new professional and academic field in the 1950s. International economic development, like global transportation, is a heterogeneous assemblage that crosses political borders, institutions, and technical systems, linking communities around the planet. In the idiom of the dominant development paradigm of the day, modernization theory, infrastructure development was seen as establishing the capital base necessary for economic "take off" via modern industry, trade, and agricultural production. Therefore, in the 1950s, the World Bank primarily made loans for infrastructure construction in the belief that economic growth depended on public investment. The state was considered central to this effort, but the governments of so-called underdeveloped countries in Latin America, Africa, and Asia were seen as lacking the requisite capacity or capital to build roads, airports, and water systems.[24]

Global infrastructures, like all infrastructures, expand and contract, creating winners and losers in the process. For example, due to the network-wide transformations introduced through the shipping container in the late twentieth century, some ports rose to prominence and others faded. Because the container could be transferred across systems without loading and unloading, the manual labor of longshoremen lost value and skilled equipment operation became essential, transforming labor relations and reducing union power. Similarly, in economic development, new theories and paradigms materialized across "underdeveloped" landscapes worldwide as roads, rural cooperatives, extension agencies, pipelines, and public health clinics. Then, when the project funding dried up (or was appropriated) or the development paradigm changed, those infrastructures would be reconfigured to fit with the new ideology or left behind.[25]

Infrastructures cross scales and provide a means of understanding their relations.[26] Thus, as explained above, global infrastructure doesn't so much "impact" local, regional, and national projects as forge relationships of interdependence with them. As infrastructures jostle for territory, they produce environments, a subject rarely discussed in social and historical studies of technology.[27]

Infrastructures, Territories, and Environmental Conflicts

The transportation infrastructure constructed around the Chagres River came into conflict with rural development infrastructure assembled by the Panamanian state. Both projects were global, in a sense, but linked to different networks. The Panama Canal developed with a network of ships, ports, maritime routes, logistical procedures, trade agreements, tractor-trailers, railroads, labor unions, and shipping containers. Meanwhile, Panama's rural development infrastructure linked campesino farmers and their fields to a network of bilateral and multilateral development organizations, traveling scientific expertise and agricultural technologies, and political currents of agrarian reform. The competing infrastructures were shaped by these linkages in ways that formatted environments and social expectations on the isthmus, giving rise to distinct ethical and political spaces.[28]

Scholars in a number of fields have conceptualized infrastructures as territorial. For example, William Cronon's study of how Chicago developed by metabolizing energy and resources from its hinterlands has been influential in anthropology, geography, history, and science and technology studies. In this literature, the city is presented as a sprawling, constructed organism that is defined by economic, environmental, and human traffic rather than political boundaries. This traffic moves via infrastructures that circulate people, information, energy, raw materials, and finished goods between urban and rural landscapes, binding them together. These "paths out of town," as Cronon calls them elsewhere, are useful to study because they mediate relationships between landscapes across space.[29]

The Panama Canal is not a city, but a state-managed engineering and administrative project. Extending a tradition of geographical scholarship on state territoriality, a growing body of work conceptualizes the state itself as a material entity that is defined through land management, sanitation, and public works projects.[30] Modern territorial administration, as sociologist Chandra Mukerji argues, is "based on 'works' rather than 'words,'" emphasizing the operation of political power through technical expertise rather than sovereign authority.[31] By transforming the landscape and enrolling nature in politics, the state extends its power over space and also increases its vulnerability to nature, because the government can then be assessed by its responses to floods and droughts.[32] To conceptualize infrastructure as territorial politics is to emphasize the materials that provide the state with its obduracy and attend to how territorializing projects are bound up with nature, competing projects, and lived realities. I build upon this scholarship by foregrounding the extension, retraction, and entanglement of two infrastructural projects purposefully organized to produce different landscapes around the Chagres River.[33]

The Panama Canal: Water and Global Connection

Empire is a territorial project by definition. It depends on the establishment of outposts far from home and the expropriation of land, often by coerced consent. The imperial state extends its authority through military interventions; the territorial politics of environmental management, sanitation, public works, and engineering; and the intimate, cultural, and biopolitical realms of categorization and moral policies.[34]

The Canal Zone, established through the Hay–Bunau-Varilla Treaty or Panama Canal Treaty of 1903, was emblematic of the linkages among empire, infrastructure, territory, and everyday life. The treaty granted the United States the "use, occupation, and control" in perpetuity of a ten-mile-wide strip of territory—more than five hundred square miles bisecting the isthmus at its narrowest point.[35] Like transportation enclaves on the isthmus before (walls around colonial Panama City) and after (walls around the free trade zone in Colón today), physical and political barriers facilitated the passage of people and goods between the oceans.[36]

The political boundaries of the Canal Zone were not fixed. In addition to near-sovereign powers within the enclave, the treaty gave the US government the authority to expropriate additional lands and waters deemed necessary for the "construction, maintenance, operation, sanitation, and protection" of the canal. Therefore, the Canal Zone was both an infrastructural and territorial project defined by an open-ended definition of what was "necessary" for canal purposes. The search for more water for the lock canal was central to the expansion of the space administered for transportation purposes. Formalized in the treaty clause granting the US government "the right to use the rivers, streams, lakes and other bodies of water within its [Panamanian] limits for navigation, the supply of water or water-power," the canal's control of regional water resources continues under Panamanian administration in the twenty-first century.[37]

Between 1904 and 1979, the US government administered the Panama Canal and governed the Canal Zone, presiding over the lives of tens of thousands of people, both its citizens and those of other nations. After new canal treaties were signed in 1977, the US government began the transfer of the control and administration of the waterway and enclave to the Panamanian government. The transfer process began in 1979 and concluded at the end of 1999. Since the beginning of the millennium, a national institution, the Panama Canal Authority, has administered the waterway and continued the engineering and territorial legacy of its North American precursors. Among the contradictions that this institution faces is the administration of nearby rural areas inhabited by campesino farmers that settled around the Chagres River during a period when the Panamanian state had a different set of development priorities.

Government of Panama: Roads and National Integration

Panama's national motto is *"Pro mundi beneficio"* (for the world's benefit) and the motto of the former US Canal Zone was "Panama divided, the world united." By opening a global waterway—an infrastructure for maritime connection—the canal builders cut the isthmus and nation of Panama in half. Since then, the country has grappled with how to integrate the so-called "two Panamas": an urban transit zone and rural agricultural interior.[38] As early as 1920, the government of Panama tried to do this by prioritizing the construction of roads that extended from the cities of the transit zone across the rural interior. If the image of a strait between the oceans—a big ditch, so to speak—captivated the imagination of foreigners on the isthmus, then Panamanians of all types during the early twentieth century were captivated by visions of paved highways and rural "penetration roads" that promised to integrate the national territory.

The Panama Canal's transportation infrastructure and Panamanian government's rural development infrastructure were both inscribed with modernist ideologies. Roads and canals were built to conquer nature and release the energies of land and water, circulate new ideas, and facilitate development. For the Panamanian government, road construction was central to the "Conquest of the Jungle" development program that channeled colonists to forested frontiers to foster economic integration and the consolidation of state power at its periphery. Similarly, the construction of the Panama Canal was understood to be a conquest over nature to circulate resources, capital, and people between the seas.

The two infrastructures were at cross-purposes due to their orientations. The canal was designed to move ships north and south between the oceans. For Panama, by contrast, the road system was designed along an east–west axis to circulate crops, people, and political influence between the terminus cities of Panama City and Colón and the rural interior. These distinct orientations were a source of frustration for many Panamanians, because the north–south transit zone effectively established a physical and political barrier to east–west connection. That buffer, in turn, made it harder for Panama to develop the kind of integrated national economy many wanted: a land communication network, rather than a waterway.

As explained above, neither US nor Panamanian territoriality was defined exclusively on the isthmus or even the continent. Rural development programs in Panama were often linked to and funded by consortia of international banks, foreign development agencies, and other bilateral and multilateral development institutions. As such, Panamanian infrastructures were bound up with changing theories, ideologies, and imaginaries of economic development from the United States, Europe, and South America that traveled across Panama, much like freighters. Development projects extended to the headwaters of the Chagres River, which was a target area for state institutions in the 1960s and early 1970s.

Infrastructure, Power, and the Panama Canal Expansion

The Panama Canal's aquatic infrastructure and Panama's terrestrial infrastructure have both been "global" in their connections, but not equally so. The Panama Canal—which transportation geographers call a chokepoint—exerts a large influence on commercial networks, while Panamanian agriculture exerts comparatively little. Chokepoints are defined as locations that, due to key physical characteristics (especially depth, width, and navigability), limit circulation capacity and force traffic to converge. Chokepoints are an important phenomenon in shipping, where they can be produced by the limits of natural passageways like straits or engineered waterways like the Suez Canal and the Panama Canal. In the case of Panama, the relevant physical characteristics are both natural and constructed through the sedimentation of centuries of transportation projects across the same strip of land. Chokepoints like the canal are valuable because they cannot be bypassed without significant costs or delays and passage is regulated (this is why the Panama Canal Authority can charge high tolls).[39]

As a chokepoint, the lock canal design is embedded in a global infrastructure that it shapes but does not control. For example, the lock dimensions established a transportation standard—the Panamax ship—that shaped vessel design, logistics, port facilities, and the geographies of global commerce. A Panamax ship maximizes one or more of the usable dimensions of the canal's lock chambers (965 feet long by 106 feet wide by 39.5 feet of draft). By the 1980s, however, shipping companies began to purchase more of the colossal post-Panamax ships too large to fit

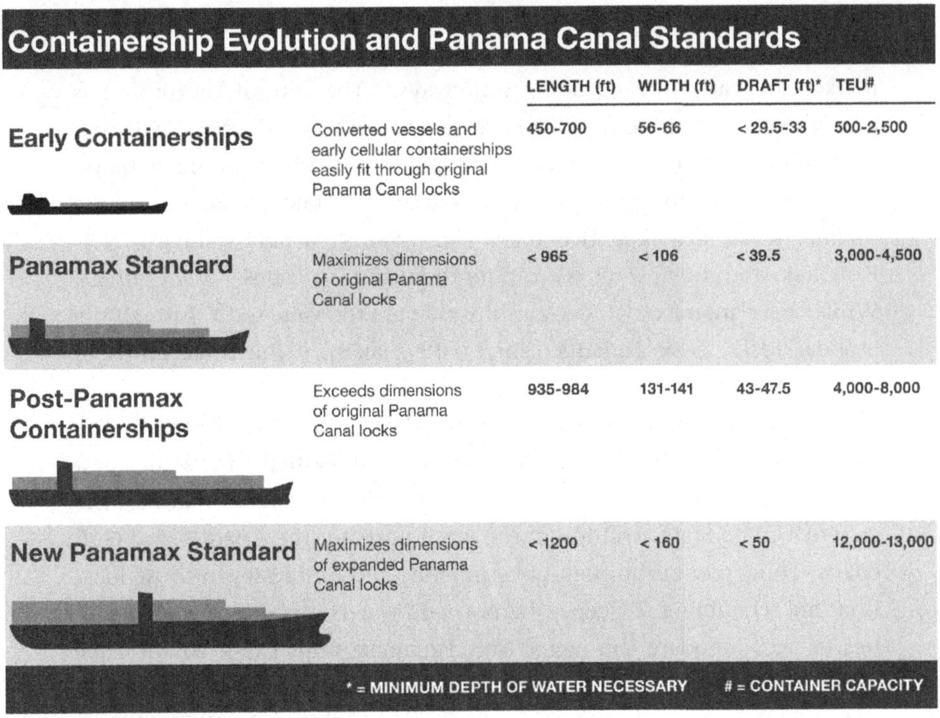

Containership Evolution and Panama Canal Standards

		LENGTH (ft)	WIDTH (ft)	DRAFT (ft)*	TEU#
Early Containerships	Converted vessels and early cellular containerships easily fit through original Panama Canal locks	450-700	56-66	< 29.5-33	500-2,500
Panamax Standard	Maximizes dimensions of original Panama Canal locks	< 965	< 106	< 39.5	3,000-4,500
Post-Panamax Containerships	Exceeds dimensions of original Panama Canal locks	935-984	131-141	43-47.5	4,000-8,000
New Panamax Standard	Maximizes dimensions of expanded Panama Canal locks	< 1200	< 160	< 50	12,000-13,000

* = MINIMUM DEPTH OF WATER NECESSARY   # = CONTAINER CAPACITY

Figure 1.7
Image by David Kuhn, used with permission.

through the canal. They could only travel busy routes between deep harbors like Hong Kong and Los Angeles.[40] Due to containerization, it became possible for shipping companies moving goods between Asia and the eastern United States to avoid the canal by routing them from deep water Southern California ports over the US "land bridge" of east–west highways and railways. Thus, in the world of the container, the canal is linked not only to a water world of maritime routes, ships, and ports, but to the US interstate system, trucking regulations, and fuel prices. Due to such shifts in transportation geography and technology, Panama Canal administrators have been under pressure for decades to modernize the aging technical system.

Panamanian voters approved a 5.25 billion dollar Panama Canal expansion project in a tense national referendum in 2006.[41] Construction began in 2007 and is expected to be completed in 2016. The expansion has been hailed as a global transportation "game changer" because it is expected to redistribute market share among ports and transform import and export markets for cargo moved along waterways.[42] The core of the project is a new expanded shipping lane with larger locks and deeper navigation channels to accommodate larger container ships. Whereas the standard Panamax ships carrying up to five thousand containers can transit the original locks, the expanded locks will establish a new standard, New Panamax, and move ships with more than thirteen thousand containers. While post-Panamax includes all ships that are too large to fit through the original locks, New Panamax ships will maximize the dimensions of the expanded locks (figure 1.7).

The canal expansion will ripple across the world of the container. The most important shift, some analysts predict, will be that the Pacific coast ports that receive the bulk of post-Panamax ships from Asia bound for the eastern United States will lose business to ports on the Atlantic and Gulf coasts. Thus, port authorities, governments, and businesses in those locations are scrambling to deepen harbors and waterways and upgrade facilities to accommodate the larger New Panamax ships expected to arrive from the Pacific via Panama beginning in 2016. Major projects are proposed or underway at nearly every major Atlantic port in the United States, including Miami, Savannah, Boston, Charleston, Houston, New Orleans, New York, Philadelphia, Jacksonville, and Corpus Christi. Meanwhile, ports in Colombia, Peru, Costa Rica, Jamaica, the Bahamas, Cuba, and Panama also have projects in the works.[43] Due to the vast reach and complexity of global transportation infrastructure, the Panama Canal expansion is expected to alter traffic patterns at railway hubs in Kansas City, create traffic jams on Texas interstates, and make US grain, Colombian coal, and Brazilian soy exports more competitive in Asia. What is often missing in discussions of the Panama Canal expansion, however, is attention to the transformations it will entail on the isthmus. In order to understand this global infrastructure, we need to better understand the politics and ecology of the canal itself.

Map of the Book

In this book, I conduct an infrastructural inversion of the Panama Canal. This research and writing strategy, which was developed by Geoffrey

Figure 1.8
Future site of Madden Dam on the upper Chagres River, 1931. *Source*: National
Archives at College Park, RG 185-G, Box 9, Vol. 17.

Bowker, begins with the premise that to understand how modern infrastructures inform social organization, we should turn them "upside down"
to examine the background work that is often taken for granted.[44] Rather
than focus on politicians, inventors, social movements, or cultural groups,
the approach foregrounds how relationships of interdependence have been
forged among technical networks and standards, social forms and conven-
tions, political regulations and bureaucracies, knowledge production,
and—I would add—landscapes. It is within these relationships of interdependence, I suggest, that environmental issues become problems.

Part 1, "Headwaters," sets up the overarching problem to be analyzed.
The section focuses on the emergence of widespread concern about water
scarcity in the region around the canal: a problem that was blamed on the
supposedly "spontaneous" settlement practices and patterns of campesino
farmers. It highlights how a cultural land cover category called *rastrojo*
emerged as a site of politics in the 1980s. *Rastrojo*—defined by farmers as
agricultural fallow and canal administrators as watershed forest—was
contested because it was situated at the intersection of two infrastructures

built to serve different populations and organize landscapes in distinct ways. The section concludes with a chapter on how the history of the watershed—a region that had never existed as a social or political unit—was created through historical land cover maps. The new environmental history of the watershed (a declensionist narrative) gave rise to specific visibilities and invisibilities, assigning rural people a new past to live by.

Part 2, "Floodplains," focuses on the construction and expansion of water infrastructure by the US government across the historically sedimented transit zone around the Chagres River. The section begins with an analysis of the historical construction (geographical, infrastructural, institutional, and social) of a transit zone across the river basin through interoceanic routes built before the twentieth century. Departing from a popular discourse emphasizing Panama's "natural advantages" as a transportation zone for global commerce, the chapter shows how isthmian society and landscape have been reorganized to facilitate movement. The chapters that follow examine US government efforts to establish a transportation environment around the Chagres River, showing how water management linked engineering and territorial administration. As canal construction came to a close, the US government depopulated rural areas of the Canal Zone inhabited by farmers and former laborers. The administrative question that emerged—how to manage depopulated areas—had implications beyond land use, because it raised concerns about the scope of US territorial ambitions in Panama. Were administrators simply operating a canal for navigation or were they building an autonomous imperial enclave? The parameters of the canal project were under negotiation in Washington, D.C., Panama City, and worked out on the ground across the Zone's fields, forests, rivers, and lakes.

Part 3, "The Interior," focuses on the construction, expansion, and retraction of the Panamanian government's rural development infrastructure across the country's interior. As commerce through the canal increased, the interior was understood to lack roads, capital, and expertise, so the government—often with the support of foreign capitalists and international organizations, embarked on a project to modernize rural Panama and extend the power of the nation-state. They assembled infrastructures of rural development—modern and global, yet in a different way than the canal—designed to produce new rural subjects and incentivize particular relationships with the land. Thus, the problems that arose as the Panama

Canal administrators attempted to manage rural landscapes to optimize the water supply available for shipping came into conflict with residual values and expectations.

Part 4, "Backwaters," concludes the book with a discussion of weeds to illustrate how the boundaries between the technical, social, and environmental are always porous and in flux. Through a discussion of managing invasive water hyacinth in the backwaters of the canal, it reminds the reader that infrastructures are embedded in the landscape. Without constant maintenance, they fall apart—weeds block waterways, traffic and weather breaks down road surfaces. Places that were formerly connected can be disconnected and landscapes that once appeared developed can revert to "nature" and be redeployed in new kinds of projects. Infrastructure also has a poetics that shapes how people make sense of the past, the present, and their places in the world, thereby producing moral economies and expectations. Human emotions, feelings, and memories may seem ephemeral, but maintaining robust infrastructure depends on addressing diverse historical experiences and tensions with humility and recognizing the limits of knowledge and control.

Part I   Headwaters

Figure 2.1
The Panama Canal Authority's Peluca hydrographic station, located on the Boquerón River. Photo by the author.

Luis invited me to come back to Boquerón to work with him on his *monte* (farm plot) after my research for this book was finished.[1] So, in June 2010, I drove from Panama City to his community of less than 150 people next to a river that shares its name. Boquerón is located in Panama's Chagres National Park and is also relatively close to the country's largest cities, Panama City and Colón—some three hours from either by bus and less by car. But nestled among forested mountains with intermittent bus service, no phone service (cell or land line), and an erratic new electrical system, the community seems a world away. Approaching Boquerón, clay-red puddles fleck the gravel road as it winds past a straggle of small cinder block houses and *ranchos* (open-sided structures with palm thatch or tin roofs). Set between the Boquerón River and the road, the town center is a cluster of white cinder block buildings in a grassy clearing: an open-air meeting house, a primary school, a small public health post, and two churches.

Boquerón looks like many communities across Panama's rural interior. Given the thick forest cover, it is easy to imagine that this area is a fragment of wilderness detached from the transportation economy downstream, but the environment here is actually managed as part of the Panama Canal's sprawling water management infrastructure. One element of the landscape betrays this connection: across the river from the town center, the Panama Canal Authority's Peluca hydrographic station collects data on water flowing through the river (figure 2.1). The station, built in the 1930s as part of the canal's flood warning system, is a material artifact and symbolic reminder of the political, hydrological, and technical relationships that bind life and landscape in Boquerón to global shipping.[2] Indeed, the very existence of the surrounding Chagres National Park (established in the 1980s across what was then a rural frontier) reflects the fact that rainfall in this area drains into the Chagres River and, ultimately, floats ships through the canal.

Luis hosted me at his house when I conducted ethnographic research in Boquerón. My return visit coincided with his month-long vacation from running the community's tiny public health post. At that time, Luis and the lone primary school teacher were the only people who received salaries (though meager government ones) to work in the community. Luis, who did not graduate high school but has completed many public health trainings, is the first responder to health problems in the community;

Figure 2.2
The Boquerón River with a low water level. Photo by the author.

the closest well-equipped medical center is an hour's travel by bus and the nearest hospital is three hours away. In addition to his job, Luis serves as a de facto community leader in the constant meetings, workshops, and trainings sponsored by domestic and international organizations pursuing forest conservation and sustainable development projects in Boquerón. These projects are aimed at transforming local agriculture in ways that are aligned with the broader goal of optimizing the water supply available downstream for navigation through the Panama Canal, as well as municipal, industrial, and hydroelectric use. When he is not at the public health post or project meetings, Luis farms. He takes time off from his job early in the rainy season to do so, because the most intensive planting (April–May) and weeding (June–July) work is done during that time of year.

When I lived in Boquerón in 2008 and 2009, most adult men were smallholder farmers, who focused on staple crops such as corn, rice, beans, and tubers like *yuca* (cassava) but who also grew bananas, plantains, peppers, tomatoes, and coffee. Some households—about one-third—raised cattle. A handful of residents in their late teens and twenties commuted hours by bus to work at Colón's Zona Libre (free trade zone), ports, and large transshipment terminal, where they packed and unpacked shipping containers destined for other Central American and Caribbean countries. These jobs at the bottom of the transport service sector paid around 10 dollars per day or 250 dollars monthly. Despite the attractions of a regular job in the city, the wages were too low to start a new life in Colón, which is a hard place to live by most accounts. Moreover, the time and money spent commuting from the *campo* (countryside) to the city were so great that commuters often quit their jobs and returned to work the family *monte*.

Monte is central to the story of how Boquerón, an agricultural community, came to be bound up with global shipping. In rural Panama, the word has a dual meaning. On the one hand, *monte* refers to a place: an agricultural plot away from the primary residence where the farmer works. Thus, campesinos in Boquerón and elsewhere often say, "*Voy pa' monte*" (I am going to my plot). On the other hand, *monte* is the pivotal land cover category within the swidden agricultural system known as *roza* in Panama and, pejoratively, as "slash-and-burn" in global environmental discourse. Swidden agriculture is an umbrella term for a variety of cultural production systems worldwide in which primary forests or secondary growth are cleared—and then often, but not always, burned—to fertilize fields cropped discontinuously with periods of fallow longer than those of cropping. Within *roza* agriculture, *monte* is vegetation considered sufficiently mature for cultivation, which ranges from secondary growth in fallow to primary forest. Because *monte* becomes organic fertilizer, *roza* agriculture as historically practiced in Panama cannot exist without it, just as mechanized farming depends on chemical fertilizers.[3] As a cyclical form of land management, *roza* agriculture is at odds with protecting stands of forests for transportation because it is organized around transforming forest into farm, which—ideally—will later become forest again.

Roza agriculture is organized around both an annual calendar pegged to the seasons and fallow cycles that extend over years and decades. In Panama, the year's production typically begins at the end of the rainy season in December. At that time, farmers select the plot(s) to cultivate during the upcoming year, based on the age and thickness of the available *monte*. When the dry season arrives in January, land preparation begins by *socolando* (knocking down) undergrowth with a machete and *tumbando* (cutting down) trees—if there are any. If the *monte* is "thick," then it could take two men several weeks to clear a hectare, but far less for "thin" secondary growth. The downed vegetation is left on the ground for a month or so until it is dry enough to burn. The preparation of the land ends with *la quema* (the burning) in March or April, just before the arrival of the rainy season. When the rains come—often sporadically in April—campesinos in Boquerón plant the first crops of the year, including: rice, corn, tubers, plantains, and bananas. Between May and July, farmers weed the plots, which will be harvested as the rainy season intensifies between July and November. This agricultural calendar is, of course, idealized and the specific preparation, planting, and harvest dates vary based on location and climate.[4]

For campesino farmers, a plot will ideally be cultivated for two annual cycles and then fallowed for ten to fifteen years before it is ready to clear, burn, and replant. *Rastrojo* (stubble) is the name given to regenerating secondary growth in fallow up to the point when it is considered *monte* again, or is mature enough to be put back into production. Thus, for people in Boquerón and other rural communities, all agricultural land—past, present, and potential (i.e., forest)—can be classified in these broad terms. Yet the terms *monte* and *rastrojo* have no fixed material referents. They are defined in relation to one another within the context of a dynamic mode of production that has historically been linked to national and global processes that shape farmers' decisions about when a *rastrojo* is "mature enough" to be cultivated again.[5]

As the social, political, and ecological context of agriculture in Panama has changed over time, the meaning of *monte* and *rastrojo* has changed as well. *Monte* historically referred to primary forest or secondary growth in fallow for at least a decade, but as the ratio of farmers to available land increased and production intensified, the term was applied to fields in fallow for only five to six years. Dramatically reduced fallow periods marked the disintegration of the *roza* system and signaled the structural problems across the Panamanian interior produced by the penetration of rural development infrastructure, market integration, and ecological degradation within the context of widespread rural landlessness. These political-ecological forces pushed campesinos to migrate to forested frontiers around the Panama Canal (like the banks of the Boquerón River) during the 1950s, 1960s, and 1970s in pursuit of *tierra libre* (available land) and higher crop yields.[6]

Luis met me outside his cinder block home at 8:30 a.m. carrying a machete and wearing a *pinta'o* (a straw hat with an upturned front brim that is identified with campesino culture in Panama), a loose yellow button-up shirt with oil stains, black slacks, and knee-high rubber boots. Breathable, skin-covering, and durable, this outfit is well suited for working on a *monte*. Rubber boots are crucial because mud and humidity destroy footwear made of fabric and leather.

We planted pineapples that June morning. Farmers in the community rarely do the same work at any given time, but most of the crops had been in the ground since April and the focus had turned to *limpiando* (weeding) the fields. However, Luis had purchased five hundred pineapple *hijos* (literally, children; actually, the spiky green tops of the fruit) two weeks before and stashed them in the shade of a young banana tree to keep them from drying out. Growing pineapple is uncommon in Boquerón, but he wanted to experiment with the crop due to its market potential. Today, he explained, we will plant the *hijos* across the hillside in a two-by-two meter grid. Luis said that he would clear the *rastrojo* and that I should follow with his homemade A-level—a triangular frame made of two sticks bound together at the top and fixed two meters apart at the bottom—in order to mark the contour at two-meter intervals where we would plant the *hijos* (figure 2.3). He moved fluidly across the hillside with his machete, cutting the tall grass that had grown up since the plot was cultivated the previous year and leaving it behind in neat swaths. I followed him with the A-level, doing my best to keep up as dark earth full of decomposing vegetation gave way beneath the soles of my boots.

Figure 2.3
The author working with Luis on his *monte*. Photo by the author.

After the rows were cleared and marked, Luis walked down the first one digging holes and I followed him, dropping a spiky *hijo* into every hole and stomping down the surrounding soil to hold it in place. We repeated the same process at the next contour line we had marked, and so on, across much of the hillside. After several hours of work, we had sweated through our shirts and stopped to rest. Luis pulled a plastic two-liter soda bottle full of cool water out from under the banana tree and we trudged to the top of the hill to relax. As we sat and looked out across the surrounding valley, he described how agriculture had changed in Boquerón since his family first settled near the river in 1958.

Luis's father and grandfather spent their first decade carving plots from the forest with hatchets and machetes. This was hard, dangerous work (tumbling trees often fell the wrong way, killing farmers). After a decade of *tumbando*, they transitioned to cycles of fallow and cropping just as things began to change quickly in Panama's countryside. A military junta headed by the young, charismatic General Omar Torrijos ousted Panama's elected president in a coup in 1968. The Guardia Nacional (the new military government) extended a rural development infrastructure of roads, cheap loans, and agricultural extension to areas like Boquerón as a means of transforming forested frontiers into productive agricultural landscapes, while extending the territorial reach of the state. This infrastructure facilitated the arrival of more people, traffic, and cattle, but for many campesinos it also carried promises of modernization and development.

Luis does not work his *monte* like his father and grandfather did and many in the community still do. He attempts to farm sustainably, but these new, intensive approaches to agriculture—which emphasize building soil, rather than burning vegetation for fertilizer—require much more labor than *roza* production, which is quite demanding in its own right. He learned these techniques through the sustainable development projects that have, for decades, brought national and international experts to Boquerón to change agricultural practices scripted as part of a "backward" culture. Yet elided in this framing of the problem (culture) and its solution (training) was the historical role of the national and international rural development infrastructures and institutions in reshaping those practices.

In many ways, then, Luis's *monte* represents the ideal outcome for sustainable development: it is a multifunctional landscape that serves navigation downstream by reducing erosion and the sedimentation of the canal's reservoirs, even as it generates food and income for his family. He recognized and welcomed the fact that he and his *monte* had been transformed through conservation and development projects. However, such openness was the exception, rather than the norm, in a community where many have become cynical about environmental projects after decades without concrete improvements in everyday life.

Luis, unlike his neighbors, has a job and a salary. This is a crucial distinction. Many farmers do not participate in projects, he told me, because managers expect community members to freely contribute their time without compensation—spending long hours at meetings, when they could be working on their *montes*. For campesinos, so-called traditional agriculture provides direct benefits, while water "protected" through environmental management projects around the Boquerón River seems to mainly enrich the downstream and overseas beneficiaries of shipping through the Panama Canal. People in Boquerón are sensitive to these national and global inequities, which are often reproduced locally in the distribution of project resources across the actors involved: functionaries, urban professionals, foreign experts, and community members. In fact, Luis estimated that 10 to 15 percent of project funds stay in the community. The rest, he claimed, goes to project vehicles, gas, and salaries.

As we will see in the next chapter, attention to *monte* and its counterpart, *rastrojo*, reveals that the Panama Canal is not just embedded in the nation of Panama or even the river basin that supplies its water. Rather, shipping is entangled with cultural landscapes worked by people whose practices and expectations have been shaped by distinct histories, institutions, and infrastructures. *Roza* agriculture is, despite critics' assertions, neither a backward nor random system. It is, as anthropologist Stephen Gudeman observes, a sophisticated "technology" that integrates material implements like the machete, the labor of the agricultural cycle, and environmental knowledge.[7] And that technology produces a landscape. At the same time, *monte*'s significance exceeds the crude economic utility of grass, bushes, and trees, because it is also a potent cultural shorthand for the sedimentation of work, place, and history as landscape. In Boquerón, settlers' arrival narratives are full of fond memories of marking out and working the first parcel, and feelings of anger and frustration around policies that restricted agriculture to protect canal water. *Monte* and its place in rural Panama have been contested in recent decades as national and international actors have attempted to redefine forests as a natural infrastructure for transportation.

3 Making the Panama Canal Watershed

Figure 3.1
Canal Zone-Panama boundary marker at Agua Clara, 1955. *Source:* Smithsonian
Institution Archives RU 7006, Box 182, Panama 1955. Used with permission.

In Panama, the word watershed—*cuenca*—didn't exist.
—Stanley Heckadon-Moreno, Panamanian sociologist, 2009[1]

Francisco Ramos is a forest guard with Panama's national environmental
agency, ANAM (Autoridad Nacional del Ambiente). When I met him at
the agency's regional office in the upper Chagres River basin,[2] Francisco—
an athletic fifty-year-old in a khaki uniform with short black hair—was
sitting at his desk, but he spends much of his time working in the field.
Since 1984, when the Chagres National Park was established, he has been
inspecting land cover on farms inside the park, including *monte* plots
worked by smallholders like my friend Luis. He and other forest guards
examine the secondary growth of grass, bushes, and young trees that rural
Panamanians call *rastrojo* (stubble) in order to determine if and when land
cover is defined as watershed "forest" under government protection.

The legal distinction between potential farmland and protected forest
is five years of growth, which means that farmers are permitted to clear

Figure 3.2
Guard post at the boundary of Chagres National Park. Photo by the author.

young *rastrojo* for cultivation, but not secondary forest beyond that threshold.[3] "Before, when there was no park," Francisco told me, "farmers were only required to get permission to cut and burn. And it was easy to get. There were no restrictions on the age of the *rastrojo*. Now, they are only allowed to cut up to five years."

As we finished the interview, I mentioned that in some of the rural communities where I worked, campesinos said that they have borne the burden of environmental regulation in the park, while the government, shipping companies, and the Panama Canal Authority benefit. The claim seemed to strike a nerve. Francisco rummaged around his desktop until he found a copy of the presidential decree that established the Chagres National Park in the 1980s and read aloud, "To preserve the natural forest for the production of water in quality and quantity sufficient for the normal functioning of the Panama Canal, as well as domestic, industrial, and hydroelectric uses in Panama City, Colón, and Chorrera." [4]

He finished and then added, "The principal objective of the park is this—to produce water—it's clear. This has been a struggle for the people whose livelihoods have been restricted. Every time we have a meeting and try to do something, they say, 'We take care of the canal, but the canal doesn't give us anything.'"

"How do you answer them?" I asked.

He replied, "What I say to them is: you aren't seeing that you have a big television, an electrical generator, and a new school in your community. All of these goods come from the canal. You are waiting for someone to come and say 'Here, take twenty dollars, it's a product of the canal.' No. The benefits of the canal that you'll receive are these goods and a better education for your kids!"

He paused. "But the people don't want that, what they want is money."

"Deforestation: Death to the Panama Canal"

In the 1970s, new concerns about regional environmental issues around the Panama Canal circulated through offices and conference rooms in Panama and the United States. Some scientists argued that the deforestation of the Chagres River basin threatened shipping through the waterway. Forester Frank Wadsworth introduced these concerns forcefully and publically at the 1978 US Strategy Conference on Tropical Deforestation, cosponsored by the State Department and US Agency for International Development (USAID). Wadsworth worked for the US Forest Service at the Institute for Tropical Forestry in Puerto Rico and had, in 1977, consulted on a USAID program in Panama to strengthen the country's environmental management institutions. In a paper entitled "Deforestation: Death to the Panama Canal," he argued that deforestation by shifting cultivators—campesino farmers—altered runoff from the watershed into the canal system, depositing sediment in the upper reservoir and reducing water storage and supply. Wadsworth described the anatomy of an emerging crisis as follows:

In May of 1977, the passage of an above average number of ships, an increased use of water for hydroelectric power and the domestic supplies of growing cities, and the production of timber, food, and forage crops within the Canal watershed led to a dramatic demonstration of the limits of the capability of the water system. The surface of Gatun Lake dropped to 3.1 feet below the level required for full Canal use. Some ships sent part of their cargo across the isthmus by land, reloading it at the other coast, and certain bulk cargo shippers even abandoned the Canal, sending very large carriers around the [Cape] Horn. In 1977, this predicament coincided with a serious drought, and this was seen as a harbinger of what could soon take place every year... Deforestation and cultivation in areas adjacent to the headwaters accentuate both flood losses through the spillway and low flow in the dry season.[5]

He argued that rural deforestation might turn one of the world's most important shipping channels into a "worthless ditch." The problem, in his formulation, could not be fixed through established civil engineering approaches to water management. "Only forests," he concluded, "can restore and stabilize the capacity of the canal. Even if Madden Dam were raised, the five additional dams built, fresh water tunneled from elsewhere, and power and urban water consumption discontinued completely, the effect of continued deforestation would be inexorable. Sooner or later it would mean death to the Canal as a reliable world trade route."[6] By invoking commercial death, Wadsworth assigned the canal a new kind of life: he rhetorically reframed an engineered waterway sometimes called "the big ditch" as part of a surprisingly fragile and regional hydro-ecological system. In his formulation, the heterogeneous character of the canal's water problem demanded an integrated political and ecological solution.

How a Technical Problem Became Environmental

As Wadsworth pointed out, Gatun Lake, the canal's main storage reservoir, dropped well below its normal operating level in 1977. However, that drought was neither the first time that the water level had dropped enough to restrict navigation, nor was it the shallowest water on record. And yet, before the 1970s, canal administrators from the United States had emphasized technical solutions to water shortages that were understood as the outcome of low rainfall and heavy traffic. Wadsworth, by contrast, foregrounded regional environmental change.

The canal was less than two decades old when a long drought in 1929 and 1930 raised concerns about the capacity of Gatun Lake to provide enough water for navigation and hydroelectric power generation when the waterway reached its full traffic capacity. Concerns about future water demands were part of the case for creating a second water storage reservoir, Madden Lake, which was completed in 1935.[7] As oceangoing ships increased in number and size after the Second World War, droughts became a recurring problem.[8] In 1957, for example, US canal administrators placed draft restrictions on ore carriers and super tankers due to low rainfall; some ships had to reduce their loads and, by extension, cut profits, to transit. As water levels fell, the Canal Zone's hydroelectric plants were turned off and replaced by diesel generators in order to divert water normally used for power to navigation.[9] A drought in 1961 forced the Panama Canal Company to issue draft restrictions once again and administrators conducted studies for yet another water storage reservoir, which was never built.[10] And, in 1964 and 1965, another long dry period arrived. Even as the canal posted new traffic records, the company enforced the strictest draft restrictions to date due to shallow water.[11]

Because water shortages were blamed on low rainfall and heavy traffic, civil engineers proposed the types of technical solutions that Wadsworth critiqued: more and deeper artificial reservoirs, long tunnels or pipes to carry water from the oceans or other river basins to the canal, or restricting non-navigation water for hydroelectric power, industry, and municipal use in Panama and the Canal Zone. The drought of 1977, he concluded, was not a periodic climatic event (as it had been in the past), but "a harbinger of what could soon take place every year" due to regional environmental change.[12] This explanation of water shortage elucidated a linkage between conservation and development that mobilized both Panamanian government officials and US canal administrators, neither of whom were previously concerned with environmental protection in the transit zone. By providing an object lesson on the potential vulnerability of the waterway to processes beyond its banks, drought forced them to reconsider environmental change in terms of broader political and economic concerns.

The Institutional Politics of Water Management

Wadsworth's call for watershed management as a necessary extension to engineered infrastructure in Panama echoed historical institutional conflicts between engineers and foresters in the United States. Wadsworth was trained in the North American forestry tradition. He received a PhD in forestry at the University of Michigan and spent his career working for the US Forest Service. Beginning in the first decades of the twentieth century, the Forest Service promoted the theory that watershed forests regulate stream flow and runoff. Therefore, foresters argued, river basin deforestation increases flood level and frequency, accelerates soil erosion, and even alters rainfall, leading to impacts on electricity generation, agriculture, navigation, and commerce. Some foresters even adopted the idiom of engineering to characterize forests as organic infrastructure, describing them as "nature's reservoirs" and commercial support systems.[13]

Between 1908 and 1911, US foresters described watersheds as "natural" political-administrative regions and harnessed public anxiety about flooding to garner support for legislation authorizing the federal government to purchase forested lands in the upper watersheds of navigable rivers. They asserted that forests regulated water flows and runoff through the so-called sponge effect. This is the idea—still debated by hydrologists— that forest cover increases water infiltration by protecting soils from compaction during heavy rainfall and loosening earth through rooting. According to this argument, forests increase the soil's capacity to absorb precipitation during rainy periods and release it slowly over dry periods, arguably leading to more consistent stream and river levels. These ideas undergirded Wadsworth's claim that upstream deforestation around the Chagres made the river flow higher in the rainy season, with excess water spilled out to sea, and lower in the dry season, when liquid was badly needed.[14]

American foresters' claims about the influences of forest cover on streamflow had deep roots, dating back to antiquity in Europe, when natural historians described the impact of deforestation on springs, rivers, and rainfall.[15] In the United States, George Perkins Marsh disseminated ideas about the hydrological effects of forests in his seminal 1864 book, *Man and Nature*, which called for watershed conservation.[16] The scientific study of the effects of vegetation on climate, water, and soil was labeled "forest influences" during the early twentieth century and, at the suggestion of Joseph Kittredge, renamed forest hydrology in 1948.[17] Beginning with the first paired watershed experiment on forests and streamflow at Wagon Wheel Gap, Colorado, in 1910, a new generation of scientists used quantitative methods and analysis techniques to understand the relationship between land cover and hydrological processes—precipitation, streamflow, evapotranspiration, flooding, drought, erosion, and water quality—previously understood only through observation and anecdote.[18]

The foresters' hydrological claims brought them into conflict with the Army Corps of Engineers, which sought to manage the same water problems through technical means. The Army Corps, which dates to 1775, is the federal agency that implemented most of the large water management projects in the United States, including dams, canals, and flood protection works. The Corps' influence on the physical geography of the United States has been profound. By engineering some twenty-six thousand miles of waterways, the agency played a leading role in establishing an extensive anthropogenic hydrological network across North America.[19] The Corps has been defined by a military ideology of order and a bias toward monumental projects, often overlooking their limits and environmental consequences.[20]

The engineering versus forestry controversy turned on scientific debates about which institution's approach—technical (engineering) or environmental (forestry)—would produce more orderly rivers, but it was also a struggle for political clout and the large governmental appropriations that were channeled to water management problems. The Corps publicly critiqued foresters' arguments for watershed management and the scientific basis of the sponge effect, which threatened their hegemony over navigation and flood control funds. Nevertheless, the US Congress passed a watershed forest protection law called the Weeks Act in 1911, authorizing the Forest Service to purchase watershed lands to ensure downstream river navigability. The Forest Service ultimately managed over twenty-five million acres of forest reserves acquired under the law.[21]

The domestic drama between foresters and engineers unfolded at the height of Panama Canal construction—the US government's most ambitious navigation project to date—but, as the Panama Canal Company's technical orientation toward water shortages demonstrated, engineers dominated water management in Panama and foresters were rare. The company (which operated the waterway itself) and, in many cases, the Canal Zone government (which presided over the surrounding enclave), were controlled by engineers whose educational and professional backgrounds were entangled with the Army Corps of Engineers, even though the Panama Canal was not officially a Corps project.[22] By contrast, forestry and environmental science had little institutional clout on the isthmus. Before the 1960s, the foresters who worked in the Canal Zone or Republic of Panama were almost entirely researchers or consultants, rather than state employees. Their work had little apparent effect on governance in the Zone or Panama, where no forestry training was available.[23]

The Canal Zone's "Squatter Problem"

Frank Wadsworth's "Death to the Panama Canal" was not simply an act of representing the canal and its region in a new way—it was an effort to build global alliances around the watershed. In his presentation of the paper in Washington, D.C., Wadsworth spoke for the canal and to the US government. By framing water shortage in a manner that militarily, economically, and morally bound the United States to Panama and the future of the canal, he aligned the interests of his audience in Washington with the protection of tropical forests thousands of miles away. But, to understand why Wadsworth's narrative of the canal's "death" attracted so much attention in Panama and the United States, the paper must be situated in its historical context. Although water shortages and draft restrictions had occurred in the 1950s and 1960s, they were recast in the 1970s against a backdrop of rapid geopolitical shifts on the isthmus and rising global environmental concern. Within this context, campesino "squatters" were scripted as an environmental problem and, by extension, a threat to the canal itself.

The smallholder farmers characterized as squatters were not a new problem for Panama Canal administrators, but ascendant environmental concerns cast their land use practices in a new light. As detailed in chapters 6–8, the Canal Zone government's racialized opposition to Panamanian campesinos and black West Indians living in the rural areas of the enclave dated to the forced depopulation of the construction era. Even after depopulation, nonauthorized farming in the Canal Zone remained a persistent, if relatively minor, concern. Many of the so-called squatters lived in Panama and crossed the Zone boundary—a long perimeter that was poorly marked and difficult to patrol in rural areas—to cultivate *montes clandestinos* (secret farm plots) hidden in the forest. The Canal Zone's small rural police force had the authority to prosecute the farmers for clearing and burning forests, but often just allowed them to harvest their crops and then warned them not to trespass again. Thus, farming in the Zone was treated as a routine matter of criminal law that was, in practice, often worked out through face-to-face negotiations between police and farmers.

In 1964, bloody "revolts" or "riots" gave voice to Panamanian frustration with the imperial US presence on the isthmus and nationalist demands for sovereign control of territory. At this highly charged moment, the incursions of campesinos into the Zone took on symbolic importance on a larger political stage. Therefore, when the Panamanian and US governments began to negotiate about the "squatter problem" in the 1960s, the problem was considered territorial rather than environmental.[24] The United States sought to protect its treaty rights and territorial power on the isthmus by maintaining a consistent policy against trespassing.[25] For the Panamanian state, the contradictions that emerged as they helped the US government manage campesinos foreshadowed the problems on the horizon as Panama assumed more responsibility for the canal and its environment.

As the 1970s began, the convergence of anticolonial sentiments in Panama and symbolic importance of campesinos for Panama's leftist military government made it difficult for the two nations to resolve problems related to squatters in the Canal Zone. During a 1973 interview in Cuba, Panamanian leader General Omar Torrijos said, "Throughout the existence of the Canal, several groups of peasants conceived that the Canal Zone land belongs to them and have established small tracts of land. The US has constantly put pressure on them to leave the land, but Panama has not tolerated that pressure because it considers that they are not a menace … to the Canal."[26] Was he wrong? The Zone police cited offenders with minor infractions like cutting a tree to hollow out for a canoe or building a *bohio* (open-sided thatch structure) with a small patch of yucca, rice, bananas, and corn. Considered case-by-case, these acts of "trespassing" were far from menacing, but, collectively, the campesinos in question were considered a problem because they were migratory and largely illegible to the state.

General Torrijos echoed the boundary-crossing acts of campesinos in his political rhetoric. He famously said, "I don't want to go into history; I want to go into the Canal Zone." Yet, for the Panamanian government, gaining control of the Zone meant playing an increasingly active role in keeping campesinos out. In 1974, for example, General Torrijos cooperated with US officials by relocating farmers evicted by the Zone police.[27] An American embassy communiqué noted the delicacy of the situation:

Nationalistic rhetoric as to the rights of these Panamanian citizens to reside in "Panamanian territory" [i.e., the Canal Zone]—which was so evident a few months ago in the Panamanian press—has been non-existent recently. This GOP [Government of Panama] cooperation promises to reduce, through probably not permanently eliminate, the vexing squatter problem. … While high GOP officials professed understanding of US concern and of the fact that deforestation of certain Canal Zone areas was not in Panama's interest, politically it was difficult for GOP to appear to side with US against its own people.[28]

Yet, by 1975, the Guardia had the future control of the canal in its sights and was "taking a very tough position, including the burning of unoccupied *bohios*, the destruction of some cultivated plots, and the issuing of citations."[29]

In September 1977, General Torrijos and President Jimmy Carter met in Washington, D.C. and signed treaties guaranteeing that the United States would transfer the canal and Canal Zone to Panama by the end of 1999. The treaty was the culmination of decades of intense negotiations too complex to detail here, but two points are salient with regard to the canal environment.[30] First, the treaties formalized the environmental primacy of the transport sector across the watershed by stipulating that Panama "take the necessary measures to ensure that any other land or water use of the Canal's watershed will not deplete the water supply necessary for the continuous efficient management, operation or maintenance of the Canal,

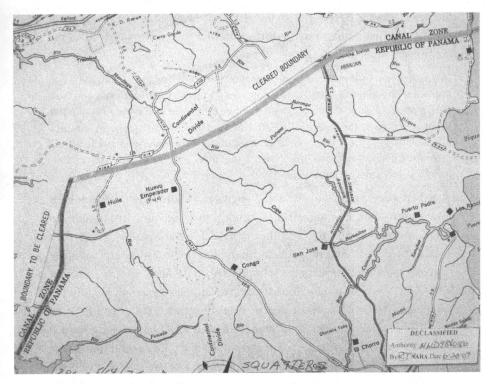

Figure 3.3
The Canal Zone border was "cleared" of squatters, 1975. *Source:* National Archives at College Park, RG 185, Entry 168, Box 25, Meeting of May 14, 1975.

and shall not interfere with the water use rights of the United States in the Canal's watershed."[31] Second, the treaty established a Joint Commission on the Environment composed of representatives from both countries and charged with overseeing environmental issues related to treaty implementation, particularly water supply and watershed concerns. As a result, joint US-Panama "antideforestation" measures including aerial and ground surveillance, education, squatter relocation, and reforestation were initiated across the watershed.[32]

Institutionalizing the Watershed

The 1970s and 1980s were a time of rapid geopolitical change on the isthmus and a period when both watershed management and tropical forest conservation were ascendant on the global environmental agenda.

Stanley Heckadon-Moreno, a sociologist who played a key role in Panama's early watershed management efforts, said that the concept arrived in the country via foreign institutions:

In Panama, the word watershed—*cuenca*—didn't exist. People knew about the canal. But when one spoke about a *cuenca*, nobody had the slightest idea what you were talking about. ... I think the word began to come into vogue in the 70s and definitely in the 80s, used by institutions like CATIE [Center for Tropical Agronomy Research and Teaching] in Costa Rica. ... The concept of using the watershed as a [political] geographical unit—not a country, not a province, or a state or a *corregimiento* [county]—but a river. That was new.[33]

Expert communities in Panama—engineers, hydrologists, geographers, and others—used the watershed concept, but it was not in popular circulation. Wadsworth collected the material for the essay in 1977 while consulting on a USAID program to strengthen the technical and administrative capabilities of Panama's impoverished natural resource agency, RENARE (Dirección Nacional de Recursos Naturales Renovables).[34] By the time he arrived, a number of experts from the United States and Panama were thinking about watershed management. USAID had already funded the research of Dr. Curtis Larson, an agricultural engineer, who found that deforestation in the watershed for cultivation and pasture increased the sedimentation of the canal and reduced its water storage capacity.[35] Meanwhile, beyond the Canal Zone, RENARE was collecting basic meteorological, hydrological, soil, and social data for analysis, map-making, and prospective watershed management.[36] But early US and Panamanian efforts were uncoordinated, a situation that began to change around the time of the treaty.

Between 1978 and 1983, the first integrated Panama Canal watershed program—funded by a 10-million-dollar USAID loan and a 6.8-million-dollar Panamanian contribution—was implemented to establish coordinated regional management to deal with the contradictions inherent across a space historically under dual jurisdiction. The program was also designed to increase local environmental awareness and "incorporate, to the extent possible, the watershed's population into the resource management conservation process."[37] Watershed management was conceptualized at the regional scale, but its promoters recognized that forest protection ultimately depended on changing the consciousness and behavior of rural people at the scale of the community and the *monte* farm plot. In contrast to rural development projects that sought to transform tropical forests into productive modern farms, the watershed projects of the 1980s aimed to incorporate rural people and landscapes into global transportation infrastructure.

Watersheds are often presented as "natural" administrative units, but the intellectual and political genealogy of this concept only dates back a few centuries. Scholars believe that French geographer Philippe Buache conducted the first river basin survey and introduced the idea of the topographical unity of the drainage basin before the French Academy of Sciences in 1752.[38] Over the next fifty years, river basin cartography became an established scientific practice and the word "watershed" entered English from the German *wasserscheide* (water-parting) early in the nineteenth century.[39] Not until the end of that century were watersheds—increasingly mapped and, thus, legible—administered as political units.[40]

Concerns about climate change, erosion, and flooding prompted the first localized prohibitions on forest clearing in catchments by the sixteenth century, if not earlier, in Europe.[41] In 1860, the French state pursued one of the earliest known national efforts to control water through watershed management by mandating afforestation—planting trees on land that was not recently forested—in hopes of mitigating flooding. State watershed management was, from the start, a redistributional project framed using the language of moral crusade. Drawing on French forestry, Napoleon III pledged that "rivers, like revolution, will return to their beds and remain unable to rise during my reign." Floods were reduced, apparently, as the rural peasants deprived of land tenure were impoverished.[42]

Early visions of watershed management were more democratic in North America for settlers of European descent, if not indigenous peoples. John Wesley Powell, the nineteenth-century explorer and cartographer of the American West, is the pioneer of US watershed management. He conceptualized that region as a series of river basins and recommended—unsuccessfully—that watershed boundaries guide settlement policy and ultimately contain self-governing populations charged with managing natural resources collectively.[43] Therefore, watershed boundaries do not determine a political approach. They have been mapped onto both authoritarian and democratic governmental visions.

The Tennessee Valley Comes to Panama

The modern exemplar of a watershed authority that transcends traditional political geographies is the Tennessee Valley Authority (TVA). Launched in 1933 in an area devastated by the Great Depression, the TVA is a public corporation organized around the Tennessee River watershed, which includes parts of seven states. Its multipurpose mandate includes navigation, flood control, power generation, and economic and social development.[44] The TVA managed a bounded watershed, but its model traveled to Europe, Asia, and Latin America. After the Second World War, it became a prototype for regional planning initiatives in countries only recently defined as "underdeveloped."[45] The model arrived in Panama in the 1980s as the government struggled to manage the canal watershed.

Colombia, Panama's neighbor, was among the first countries to adopt the TVA model of integrated watershed management and development in its Cauca and Magdalena River basins. The Corporación del Valle del Cauca (Cauca Valley Corporation) was established in 1954 with the assistance of the TVA administration.[46] Like the TVA, the Cauca Valley Corporation employed a public corporation model and had multiple mandates, including: transportation, flood control, irrigation, power generation and distribution, environmental management and conservation, and the promotion of social and economic development.[47]

The Tennessee Valley model came to Panama via Colombia during a series of meetings convened in 1985 and 1986 by a national task force on the Panama Canal watershed.[48] The meetings, which brought together government officials, scientists, military leaders, international experts, and functionaries, were a pivotal moment in the institutionalization of the watershed as a Panamanian—rather than US-controlled—administrative space. The Ministry of Planning and Political Economy sponsored the meetings with the support of the Guardia Nacional's new Department of Environmental Protection. The fact that the military, now under the control of the ruthless dictator Manuel Noriega, had an environmental division signaled that conservation had taken on new value.

The meeting organizers and participants recognized that conflicts between state institutions with different mandates—particularly agricultural development and environmental protection—stood in the way of the goals of watershed management and thus saw institutional coordination as a goal. The organizers had contacted the TVA for assistance with an integrated watershed management model and were referred to the Cauca Valley Corporation.[49] A representative from Colombia, Alberto Patiño Mejía, presented on the corporation's history, objectives, and programs. He emphasized that rural development programs were central to the management of the Cauca River watershed.[50] However, Panama's Chagres River basin was different because of the hegemony of transportation.

"The Canal," an early evaluation of the USAID watershed management program explained, "represents Panama's major industry and is at the heart of a complex system of support and service industries ... the project benefits Panama's major industry and its work force."[51] Within this framing of the national economy, forests were understood to "produce" water, the lifeblood of the canal and the transport service economy (a problematic metaphor legally encoded in the presidential decree that created the Chagres National Park). But forests could not be protected without also dismantling the existing rural infrastructures and institutions that had previously been assembled to reorganize the same forested landscapes to "serve" different communities in different ways.

Indeed, USAID's phrasing—*incorporation of the watershed's population into the resource management process*—signaled a dramatic shift in who was assigned responsibility for canal water. As forested landscapes were assigned an infrastructural function (water provision), their inhabitants were simultaneously charged with a new responsibility (forest protection). However, the implementation of watershed management would prove difficult in practice because its environmental goals were in direct conflict with those of an established infrastructure that supported agricultural development.

The Panama Canal watershed was, like the TVA before it, a new way of governing space. As landscape architect Jane Wolff writes about the TVA, "The agency's purview was determined not by political boundaries or local identities but by a geomorphological condition—the watershed of the Tennessee River and its tributaries. Before the TVA, physical geography had never been used to specify the powers and limits of an American agency."[52] In practice, the heterogeneity of the watershed's human geography was not easy to manage. Its inhabitants were rich and poor, urban and rural, and white and black. They occupied distinct social and economic worlds.

TVA administrators made the case for integrated watershed management through a narrative of environmental decline, much like Frank
Wadsworth's "Deforestation: Death to the Panama Canal." In *TVA: Adventure in Planning* (1940) the preeminent British scientist Julian Huxley
wrote that the watershed's virgin forest was cut by loggers, then farmed
until the topsoil washed away. Land seemed abundant, he wrote, so farmers
worked plots for only a short time before abandoning them to pasture and
moving on.[53] The TVA's solution to this problem was to reconceptualize
landscape at a regional scale through public works like hydroelectric
dams with rural roads, recreation facilities, and land restoration programs.
Through infrastructure, the Authority sought to produce a regional landscape that appeared both integrated and coherent—a goal pursued through
the transformation of physical geography and cultural land use practices.
Integrated watershed management in the US Southeast was thus, as Wolff
notes, simultaneously progressive in its vision and brutal in its historical
erasures.[54]

New infrastructures make older ways of life difficult to maintain.[55] The
establishment of the Panama Canal watershed as an administrative region
depended on campesinos accepting new environmental responsibilities
incompatible with those that the Panamanian state had historically
assigned them: agricultural colonization and development. Given the
rapidity of this shift and tensions between old and new state plans for
the region, it proved difficult to convince rural people that the forests
they lived and worked in were not theirs alone, but part of a regional
hydrological support system for shipping. Watershed managers encountered, at every turn, a rural development infrastructure established by the
state (roads, agricultural cooperatives, extension agents, agricultural loan
programs) that encouraged the agrarian land use practices they considered
economically and ecologically irresponsible. Watershed management thus
entailed retracting the existing physical and institutional infrastructure,
while reworking the campesino moral economy and expectations of development that had accreted around it.[56] Administrators recognized that the
success of watershed management was contingent on enrolling forest
guards with the capacity to align the diverging interests of state institutions
and rural social worlds.

Lucho's Story: Watershed Management on the Ground

Lucho grew up farming near the headwaters of the Chagres River, but before he became a forest guard, he had never heard anyone use the word *cuenca* (watershed) to describe the area. He moved with his family from Panama City to settle on the banks of Alajuela Lake in 1958, when the canal reservoir was called Madden Lake and controlled by the United States. Like many settlers arriving at the time, Lucho, still a teenager, dreamed of farming his own land. He wanted to work independently, not as an *empleado* (wage laborer). One day in 1975, he was working on his *monte* when he received a note that Colonel Rubén Darío Paredes, the well-known Guardia Nacional leader who was then minister of agriculture, wanted to meet with him. [57]

When they met, according to Lucho, Paredes said, "You've been recommended as a man who is not afraid of anything. We'd like to give you a job: we want you to keep the hand of the campesino from destroying the watershed." Lucho, unclear about what this meant, asked, "What is the watershed?" Paredes said, "The watershed is all of this area that drains into the lake." Lucho recalled that he then told Paredes, "I'd like to do it, but I have to talk with my wife, my first child is on the way." Paredes offered him fifty dollars every two weeks, but Lucho countered, "I'm not going to abandon my land for fifty dollars, colonel. I've got an old mother, an old father, a brother—we can't live off of that much money. I'm my father's right hand." According to Lucho, Paredes increased the offer to include free education in natural resource management. He had no interest in the environment, per se, but was interested in the promise of education.

Lucho was thus enrolled in the creation of a new natural infrastructure for shipping.[58] Paredes had mobilized the promise of career opportunity to convince him to put down the machete and become a forest guard defending the watershed from the machetes of other campesinos. RENARE—the predecessor of ANAM, the current environmental agency—recruited an initial group of forty-six forest guards to patrol the watershed.[59] The others, like Lucho, were identified by officials as leaders in rural communities. Urban watershed managers hoped that guards familiar with the area and its people would facilitate local cooperation. The forest guards' first project was to survey the human population living within the watershed. They spent three years—1975 to 1978—conducting surveys in order to establish a demographic baseline for the watershed, effectively assigning inhabitants to the new region for the first time.[60]

Regions are not self-evident. They are human creations, demarcated and brought into being in the face of a multitude of alternatives for producing space.[61] Thus, the enumeration of the watershed's population did not make them its inhabitants, because they did not inhabit space in ways that were aligned with its geohydrological contours. For campesinos, the important spatial frames were the household, the *monte* plot, and the routes that connected them to markets and other sites of social and economic significance.[62] Forest guards were charged with region-making on the ground by translating extralocal concerns about forests, water, and the canal in communities. The enrollment of campesinos in watershed management was emotionally and physically demanding for the guards. Rural communities pushed back against conservation. And, paradoxically, the geographical features that made the upper watershed area so valuable for water provision and storage—heavy rainfall, dense forest, and a lack of roads—also frustrated the guards' efforts to move across the landscape and patrol land use.[63]

Forests Politics between Two Infrastructures

During the 1979 to 1999 transfer of the Canal Zone from the United States to Panama, the Panamanian government established two national parks in the Chagres River basin, both oriented toward "producing" water for the canal, among other uses. However, the areas that became parks had distinct histories of settlement and land use. Soberanía National Park, created in 1980 on the east bank of Gatun Lake, protected around eighty square miles of largely uninhabited forest from the former Canal Zone. By contrast, Chagres National Park, established in 1984 across nearly five hundred square miles of the river's headwaters, protected an area that had been actively developed by the government of Panama in prior decades. The enclosure of the second park brought campesino communities and the state into conflict around both the meaning and materiality of forests.

Watershed management took a more coercive turn after the Chagres National Park was founded. Armed soldiers from the Guardia Nacional's

Department of Environmental Protection—the eco-emissaries of Manuel Noriega's often brutal military—began making joint inspections with Lucho and the other forest guards, after some of the guards requested permission to carry guns in response to strong resistance and threats of violence in rural communities. Together, the forest guards and soldiers supervised critical watershed sites by land, water, and air to ensure compliance with environmental laws.[64]

A potent symbol of campesino culture, the machete was officially recast-from a technology of development to a threat to the national economy. In 1987, another new environmental law, Forest Law 13, legally defined watershed vegetation more than five years old as "forest." The law meant that campesinos could be fined not only for clearing primary forest, but also fallow land previously in agricultural production that was covered by secondary vegetation. In cases remembered with anger decades later, people living in the national park were jailed or had their machetes and hatchets confiscated. Strict enforcement provoked outrage in rural communities, which became more hostile to the guards.

When watershed managers made the case for forest protection in urban and institutional settings, the referent—land covered with trees—seemed obvious. However, cooperation was more difficult in the charged encounters between forest guards and campesinos. As explained in chapter 2, the forests known and worked by campesino farmers were not a fixed object—a green area on a map—but a dynamic element of their *roza*, or swidden, agricultural system. For these agriculturalists, the longer a *rastrojo* grows before it is cleared, the more nutrients available for the next crop on that land. Consequently, farmers weigh the maturity of a *rastrojo* against pressures and incentives to put land back into production as they make clearing and planting decisions. Or, to put it another way, land use is shaped both by the farmer's relationship with the land and the location of that relationship within a broader political ecology that constrains and shapes choices.

Many swidden farmers in Panama would choose to clear young *rastrojo* only in the absence of available *monte* (primary forest or older secondary forest). However, Panama's Forest Law 13 of 1987 redefined *rastrojo* more than five years old as protected "forest," thereby encouraging shorter fallow cycles. Farmers began to clear *rastrojo* earlier (the preferred fallow period for a plot among *roza* agriculturalists is more than a decade) so that their farmland—most often owned by possessory right rather than formal title—would not fall under state protection in perpetuity. In the short term, then, the unintended consequence of this policy coupled with population growth was that because farmers reduced fallow time to maintain land tenure, the *monte* did not reach its productive potential. From a downstream perspective, shorter fallow periods may have even increased soil erosion and the sedimentation of the canal's storage reservoirs.

Conclusion: Natures, Infrastructures, and Histories

My intention in this chapter is neither to romanticize swidden agriculturalists nor to vilify Panama Canal administrators. Indeed, Panamanian *roza* agriculture may never have been sustainable in terms of its capacity to regenerate sufficient secondary forest for production without incorporating new *montes*, even in instances of low population density and market pressures. Moreover, farmers and ranchers supported by state institutions played an undeniable role in deforesting the watershed between the 1950s and the 1970s.[65] Yet, early watershed managers responded to this problem through strict forest preservation—a strategy that focused on individual behavior—rather than conservation sensitive to the structural and infrastructural realities that have conditioned campesino land use.

Since 1997, the Panama Canal Authority, the quasi-autonomous Panamanian state institution that administers the canal, has been responsible for managing, maintaining, using, and conserving the hydrological resources of the Panama Canal watershed. The canal authority, in collaboration with the national environmental agency, ANAM, emphasizes local participation in watershed management and has sought to develop a "water culture" in rural communities through consultative committees made up of local leaders from "subwatersheds" (smaller drainage basins within the canal watershed), public relations campaigns, and environmental education programs. Yet, despite these changes in approach, familiar questions of justice persist in rural areas, because the historical dispossession associated with watershed management has gone unrecognized.

The watershed emerged as an environmental problem space during a period of rapid geopolitical change on the isthmus and at a moment when the watershed concept was ascendant across global environmental and development networks. Panamanian forests *became* infrastructure through

Figure 3.4
Watershed management projects in the Chagres River basin. Photo by the author.

the multiscale organizational work of linking rural landscapes with an engineered water management system and new national and international institutions. Thus, watershed forests were purposefully constructed at great expense to serve specific social, economic, and political priorities—not unlike a dam or a highway. USAID has invested three decades and hundreds of millions of dollars in the project.[66]

Building natural infrastructure entailed—and still entails—the intimate work of forging and maintaining managerial relationships with rural people whose land use practices were scripted as a threat to global commerce. Much of this contentious infrastructural work fell on forest guards like Francisco and Lucho charged with enrolling campesinos in watershed management and translating extralocal environmental concerns. Making the watershed was subtle, yet coercive because it involved disconnecting campesinos by retracting a rural development infrastructure of roads, cooperatives, extension agents, and loan programs that was assembled to metabolize forested landscapes to different ends.

Therefore, tensions in the watershed were not simply a struggle between "global" and "local" actors over forests as a fixed natural resource. Rather, they turned on the different ways in which anthropogenic environments are constructed around and incorporated into infrastructures to serve different purposes. As the example of *rastrojo* illustrates, these landscapes do not have value in an absolute sense. Rather, they have a variety of potential capabilities that emerge in relation to particular uses.[67] When a landscape is assigned value in relation to one infrastructure or cultural system of production (transportation) rather than another (agriculture), different services become relevant (water provision rather than nutrient delivery), and the landscape is reorganized to prioritize the delivery of those services and support that system. This calls us to examine the ethics of reorganizing nature as infrastructure and to ask how systems like the canal might be designed in a manner that is more just and equitable for their neighbors.

4 Frank Robinson's Map

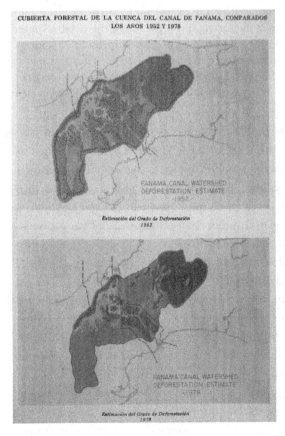

Figure 4.1
The year 1952 is the baseline for deforestation in the Panama Canal watershed.
Source: "Environmental Profile of Panama," a report written for USAID by the contractor International Science and Technology Institute, Inc., Washington, D.C.

The landscape changed rapidly as our small airplane approached Panama City. Past the forested mountains around the upper Chagres River, low-density urban sprawl gave way to older parts of the city and a wall of glass skyscrapers pushed up against the waterfront. Ships dotted the Bay of Panama waiting their turn to enter and transit the canal. I flipped through Air Panama's in-flight magazine and came across an article entitled "The Miracle of the Watershed" nestled among glossy photos of water, forest, and sky. In it, the author describes the relationships among the rural, urban, and maritime landscapes visible from the airplane:

It is in the "guts" of the canal's watershed that trees collect water, acting as sponges that absorb the rain and then slowly allowing it to flow freely even in the dry season. Here, immense trees hover over the undergrowth, protecting the riverbeds from excessive sedimentation. ... It is one of the richest of protected areas in the country and is the guardian of the extraordinary water resource. ... We are certain that it would be difficult to imagine the Panamanian economy, based on cargo container ships, without this blessing ... It is precisely this water that moves the Panamanian economy.[1]

In the article, the watershed is presented as both pristine nature and functional machine. But the region that is known and administered as the Panama Canal watershed was not always recognizable to members of the public like the passengers on my Air Panama flight. To be described as miraculous, abused, or threatened, the watershed had to be stabilized as a coherent geographical unit rather than a meshwork of people, technologies, rivers, laws, and ecosystems.

To understand the stabilization of the watershed as a region, we have to return to 1952, the year the first map of its forest cover was made. At first glance, the map made that year seems straightforward. It depicts a river basin under dual jurisdiction (figure 4.1). The Canal Zone bisects the watershed and is almost completely forested. Several strips of deforested land touch its western boundary with the Republic of Panama. To the east of the Zone, the Transístmica (transisthmian highway) between Panama City and Colón—a decade old in 1952—is the only deforestation corridor, besides minor clearing along rivers. However, what the map does not reveal is the context of its creation and how it became the definitive benchmark against which to measure regional forest cover and, in a more general sense, to reimagine the present by means of a historical artifact.

Few paid attention when the 1952 map—which suggested a regional wilderness around the canal—was created. It was not until the 1970s, when environmental anxiety grew around the waterway, that the map began to circulate among canal administrators, development experts, and state officials via national and transnational networks of conferences, publications, meetings, and projects. The 1952 watershed map did not travel alone. It was inevitably compared to either a 1976 or 1978 map depicting a basin in which the Canal Zone was still forested, but vast swaths of trees had disappeared across the border in Panama (figure 4.1).

Together, the maps supported a new narrative of regional environmental decline. Frank Wadsworth used them as visual evidence of rapid deforestation and the attendant need for environmental management in his alarming 1978 report "Deforestation: Death to the Panama Canal." "Originally covered with dense rain forest and still 85 percent forested as recently as 1952," Wadsworth wrote, referring to the baseline map, "some 250,000 acres, or 35 percent, of the Canal watersheds [sic] have since been deforested—burned for cultivation or pastureland."[2]

The map is not the territory, but it helps to bring it into existence. Harnessed to a narrative of environmental decline, the watershed maps were more than mere representations; they were spatial propositions.[3] For Wadsworth and others, reading the 1952 map against the 1976 map suggested a linear historical process of watershed deforestation caused by campesinos. This interpretation ignored the pressures and incentives from urban markets and national policies beyond the watershed that shaped rural land use within it. Over the years, this narrative—almost always illustrated with before-and-after maps—circulated widely and was interpreted uncritically, playing a key role in stabilizing the Panama Canal watershed as a region with an upstream environmental problem.

The maps were not wrong, but they could be easily misread because they divorced the watershed from its broader historical (pre-1952) and geographic (binational and urban) context. Like the perspective of Panama City seen from the window of an airplane, the map made spatial patterns visible, while obscuring the historical processes and relationships that produced them—in this case, the multiscale politics that precipitated land cover change. On the ground, noncontiguous moments and geographies are often linked in ways invisible from above.

Figure 4.2
Frank Robinson at Agua Salud, 1981. *Source*: *Panama Canal Review*, October 1, 1981.

The 1952 map became the baseline for measuring environmental change around the canal. In order to understand how the region's environmental problems and solutions were defined, then, we need to consider how the map, an artifact with its own political and cultural history, was constructed. Thus, we turn to the story of another Frank—not Wadsworth, but Frank Robinson—who made the map. He, like Wadsworth, was a man from the United States creating representations of Panamanian forests, but his understanding of the isthmus—gained over decades working for the Panama Canal Section of Meteorology and Hydrology—ran much deeper.

Robinson reflected on the watershed maps in an oral history conducted by the US government in 1982 as part of a series about the Canal Zone.

Robinson: Okay, we'll talk about the watershed first. Look at this [map] ... The green depicts more or less what is left forested, this orange area is a little boundary. This is your watershed. In the pink or reddish part is [*sic*] areas that has been cut over and the original forest gone.[4]

Later in the interview, he explained the origins of the maps and referred to how they traveled from his US institution to a Panamanian one in the 1970s:

Robinson: I made this one in 1952. When I first came to work here I was a young knucklehead and I loved the bush. And this one I made in 1976. And I made several others in between. Then RENARE [the first Panamanian environmental agency] took them after the treaty, and went ahead and produced their set from mine.[5]

When Robinson arrived in Panama in 1946 he was a twenty-year-old veteran of the Second World War from Sanford, Florida. His brother Robbie told me that their father moved to Panama first in the 1930s without the rest of the family to work for a military contractor, because he went broke during the Great Depression. Frank and Robbie spent much of their child-hoods in Sanford with their mother. At that time, Sanford was swampy and poor. Most people made a meager living growing celery or raising a few animals, not unlike rural Panama.[6]

Frank's mother and brother moved to the Canal Zone in 1945 and he arrived to join them in 1946, after the war. At first, he hated the isthmus because it reminded him of the Philippines, where he had been sta-tioned during the war; he quickly left for college in the United States.[7] Frank returned in 1952 and was hired by the canal's Section of Meteo-rology and Hydrology, or "Met and Hyd" as it was called in the Canal Zone, to collect data from the hydrographic stations distributed across the Chagres River basin. Founded in 1905, Met and Hyd was responsible for reporting on "all natural phenomena which may affect the water supply and consequent navigation of the Panama Canal."[8] Like Frank Wadsworth, who spent his career studying forests in Puerto Rico, Robin-son's journey from military service in the Philippines to collecting data along Panamanian rivers traced the far-flung US imperial network in the tropics.

Robinson was not a cartographer and his job did not entail mapping forests—or anything to do with forests, for that matter. What's more, Met and Hyd had no authority across the upper portion of the river basin under Panamanian jurisdiction. The headwaters were located in a different country.

Robinson was drawn, nonetheless, to the forests of the upper watershed. As he said in the interview excerpt above, he loved the bush. In fact,

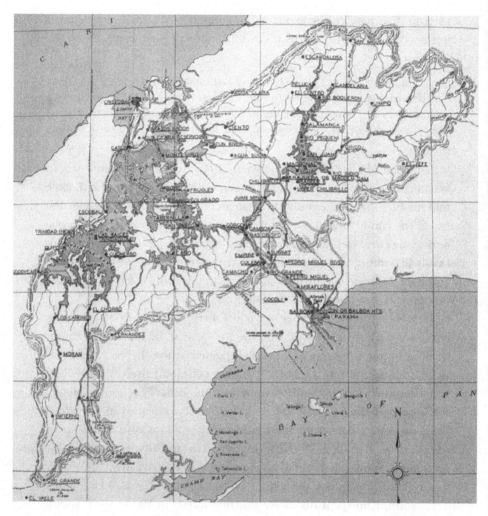

Figure 4.3
Watershed map with meteorology and hydrography stations marked, 1949. *Source*:
George Matthew, "Chagres River and Gatun Lake Watershed Hydrological Data,
1907-1948" (n.p.: Panama Canal Section of Meteorology and Hydrology, 1949). 2.

whenever his name came up during my fieldwork, the speaker (campesino, tugboat pilot, hydrologist, or natural scientist) remembered him as a "bush guy"—code for a type of American man distinguished by testing himself against tropical nature, but not directly dependent on the land to survive like a farmer. The Met and Hyd job was a good fit for Robinson because its staff, which operated out of a field office near the Pedro Miguel Locks, spent a great deal of time in the field. The canal's water management network extended to areas in the headwaters of the Chagres and its tributaries with few roads. Thus, a 1981 profile in the *Panama Canal Review* summarized the fieldwork of Met and Hyd "water watchers" in this way:

While digital recorders and the telemetering system have greatly facilitated today's data collection, many aspects of the hydrologist's and technician's jobs have not changed over the century. Personnel must still trek into the jungles to check and maintain the instruments at the stations. ... Because of the remoteness of the stations, travel accounts for about 20 percent of the branch's field activity. Four-wheel drive vehicles, horses, *cayucos* [canoes], *piraguas* [flat-bottomed boats] and even helicopters are all used to venture out into the field. The most frequently used mode of transportation besides one's own two feet, is the *cayuco*, a long narrow boat carved from a tree trunk and used by the Indians for centuries. No craft has been found better suited for navigating jungle rivers.[9]

Thus, Robinson and his colleagues—who were predominantly white, English-speaking men from the United States—came into contact with rural Panamanians while conducting fieldwork due to the transboundary geography of the infrastructure they managed and maintained. As boundary workers, their travels illuminated the continuities and distinctions between the social worlds of the Canal Zone and rural Panama.[10]

The canal company purchased canoes that "local" Choco Indians (who were actually, like the hydrologists, recent migrants to the area) carved from tree trunks up to fifty feet long to access remote stations like the Peluca Station on the Boquerón River.[11] The everyday work of measuring rainfall and streamflow was, much like farming or canal operation itself, bound to the seasonal rhythms of precipitation. During the dry season, as nearby campesinos cleared *monte* to plant when the first rains arrived, Met and Hyd employees repaired and maintained equipment at the hydrographic stations and cleared overgrown trails nearby. During the rainy season, especially the heavy rains of October and November, campesinos completed the year's harvest and field crews from Met and

Hyd spent one week per month at each hydrographic station to measure peak streamflow from a gauging car suspended above the river.

Robinson was later called the "father of the canal watershed" because he was among the first to be concerned about the hydrological effects of deforestation.[12] Although water—not forest—was the chief concern of Met and Hyd employees, the rivers and trails they traveled into the field were the same ones used by the wave of Panamanian campesinos settling the region, which made changes in forest cover highly visible. Yet there was also a politics of visibility at play.

The rivers and lakes were canal water sources under US jurisdiction, but all forests beyond the Canal Zone border were under Panamanian control. This is why Robinson's 1952 map isolated the watershed—half in Panama, half in the Canal Zone—from its political geography. During the oral history interview he conducted three decades later, he explained, "During the political days [before the signing of the 1977 canal treaties], I would only put in the watershed, because it was none of my business what was happening up here [in Panama]."[13] Freed from its political, geographical, and historical context, Robinson's map was a spatial proposition because it suggested a canal administrative region that did not exist. However, by the 1990s, its silhouette would be reproduced in project documents and newspaper articles, and discussed in community meetings and schoolrooms.

For two decades, Robinson mapped forests alone. In 1972, Met and Hyd hired Luis Alvarado, a young Panamanian hydrologist educated in Europe who was also concerned about the changes in forest cover that Robinson— his new boss and mentor—had been carefully mapping. They began to collaborate. Alvarado described their cartographic process to me in an interview: "It was done very crudely, simply with visual observation out of helicopters when we were collecting [hydrological] data ... when we were going in, you could see where new forest was being cut. We'd fly with maps and then we'd pinpoint it. No GPS or anything. You knew the river and you knew more or less where the area it was 'that looks like ten hectares' and you'd make a little circle: ten hectares. Very crude but very effective, because over ten years you could see the difference in what was happening."[14] But what, exactly, was happening on the ground?

Nobody was paying attention when Robinson—the "bush guy"—made the first forest map in 1952. But, as explained earlier, the map *became* important in relation to the maps that he made with Alvarado in the 1970s. As the Canal Zone was transferred to Panamanian control in the 1980s and 1990s, the stark before-and-after images circulated on the isthmus and overseas. Even as more accurate land cover maps were made using aerial photography and satellite imagery, 1952 continued to be the baseline year for measuring regional environmental change, because no historical spatial data on forest cover was in wide circulation except Robinson's map. Thus, the map and the year it represented have remained pivotal.

Watershed history might be defined along two axes.[15] The first axis is a chronological series of years in which the events of 1952 happened *before* the drought, water shortage, and signing of the Panama Canal treaties in 1977. But, as the watershed map illustrates, much of the environmental significance of 1952 was defined *retroactively* in the 1970s, 1980s, and 1990s due to the establishment of the watershed as a new political and administrative unit. The second axis, then, can be understood as the sedimentation of past years with new meanings and relevance. Or, to paraphrase Bruno Latour, the watershed of 1952 as it is known today does not have the same components, textures and associations as the Chagres River basin of 1952.[16] The forests mapped by Robinson alone, then Robinson and Alvarado, became the concern of a growing network.

Frank Robinson's map represented a space that, at the time of its creation, had never existed as a political or social reality. Taken out of context, it encouraged a misreading of the landscape. The heavily forested "baseline" image of 1952 suggests a nearly pristine wilderness region. Beginning with that assumption, the arc of environmental decline (from primary forest in the 1950s to degraded landscape in the 1970s) was presented as linear and projected into the future to justify political interventions. Yet, for reasons explained in chapters 6 and 7, the Chagres River basin had gone through several cycles of deforestation and forest succession during the first half of the century. This meant, for example, that many of the trees mapped near the canal in 1952 were less than forty years old, dating to the enclosure of the Canal Zone and end of the construction period.

The map was a simplification of the social, political, and environmental realities of the watershed, but, to be fair, it also simplified Robinson's more nuanced perspective. In his 1985 publication, "A Report on the Panama Canal Rain Forests," the first work of its kind, he vividly described the changes he saw over the decades in the watershed, demonstrating a knowledge of Panamanian history and culture uncommon among Zonians.[17] "The watershed," he concluded, "can be described as a giant 'scale' measuring the needs of the future against the present needs of its inhabitants ... whose livelihood depends on the destruction of the rain forests to provide land for farms and pasture."[18] My ethnographic research suggests that, before the 1977 treaties, Frank Robinson and other Met and Hyd technicians were among the few representatives of the Panama Canal that were known to campesinos in the upper watershed. Robinson visited people around Boquerón in their homes, fished with them, and hired them when possible.

The forest cover maps—especially Robinson's 1952 and late-1970s maps—obscured as much as they revealed. They became a tool for interpretation and action within a network of relationships distinct from those surrounding their production. As a justification for intervention, they would reshape the reality they represented, producing a watershed space that resembled, but also exceeded those representations.[19] Anthropologist Gregory Bateson observed that the significant moments in history are when the default settings change.[20] Robinson's maps, which highlighted spatial processes ripe for interpretation, were central to the process through which watershed "inhabitants" were given a new past to live by.

Part II Floodplains

5   Life along the River (Miocene-1903)

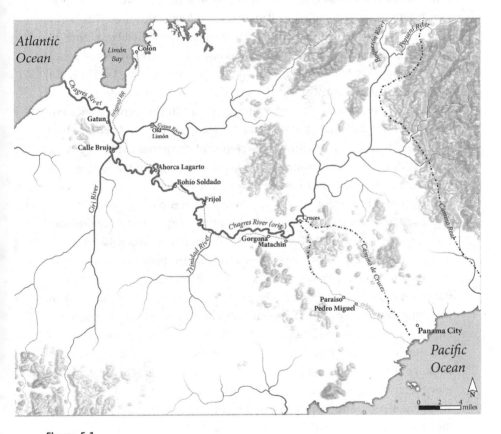

Figure 5.1
The Chagres River region in the late nineteenth century. Image by Tim Stallman,
used with permission.

The Panamanian historian Bonifacio Pereira Jiménez observed that the history of Panama is, to a certain extent, the biography of the Chagres River.[1] The existing lock canal is only the most recent chapter in that story. The land and water routes that grew up around the river carried fabulous wealth from Spanish colonies in South America and then the gold mined by US forty-niners in California. Today, the river floats containerships and tankers carrying cargo from around the world.

Geography and history are inseparable in accounts of Panama. Observers have long suggested that the isthmus's most important natural resource is not beneath the land (minerals, oil) or upon it (crops, forests), but location itself—or, as it is called in Panama, *la posición geográfica*. This is the notion that, as one historian put it, "The country was called by 'nature' to be an emporium for commerce."[2] Scholars and other writers have long used the geography-as-resource trope to frame Panama's canal. In 1914, for example, a historian wrote, "the demands of international trade automatically decreed that peace and order should prevail in the territory adjacent to the natural pathway of commerce."[3] Nearly a century later, another historian observed that isthmian development "required exploiting their remarkable geography. Panama had little coffee or sugar to sell, but, poised on a narrow strip of land between two great oceans, it had served as a crossing point for explorers and conquerors for hundreds of years."[4]

Narrated in this way, Panama's role as a transportation service provider for the contemporary global economy seems inevitable, rather than socially and materially constructed over time. Geography becomes destiny and history is reduced to the activation of the land's potential by people with the technology, skills, and capital deemed necessary to do so. The problem with this philosophy of history is not that it assigns physical geography an active role in human affairs, but that it naturalizes engineered systems like the canal, rendering the ongoing and highly political background work involved in facilitating transportation invisible and, thus, seemingly inevitable. As anthropologist Joe Masco observes, "This type of modernism encourages people to approach the invention of new technology as an inevitable part of an evolving natural world, and not as a cultural product that requires everyday decisions and infrastructural support and that produces profound cultural contradictions at every level of everyday life."[5]

To illuminate the purposeful, political, and ongoing work of facilitating navigation across the isthmus, we need to invert how Panama's historical geography is often read. Because there are no natural straits between the Atlantic and Pacific, interoceanic routes must be built and maintained by people. The oceans are close in Panama, to be sure, just as they are at points in Nicaragua, Mexico, and Colombia. Yet, those countries have no canals, even though projects have been proposed for centuries. The distinctions between Panama and less developed interoceanic transit zones in the Americas have as much to do with the history of infrastructure and governance as natural advantages (i.e., the width and topography of the isthmus). As transportation geographers have shown, the spatial structure of transportation networks is an outcome of the complex interplay of physical geography, historical contingency, and infrastructural inertia. Thus, the succession of projects around the Chagres River since the fifteenth century has produced a sedimented transportation environment in which each additional route built has depended upon (and sometimes enhanced the significance of) its precursors, while laying the groundwork for the projects that came afterward by modifying the physical, political, and social landscape.[6]

Panama's "natural advantage" and, by extension, its "geographic occupation" as a transportation service provider, must also be understood in relation to the historical accretion of capital around the Chagres River in the form of infrastructure and institutions, rather than simply the often touted physical attributes of the isthmus. Seen in this way, the reason that Panama had little coffee or sugar to sell (to use the example above) is due, at least in part, to how the built environment has been organized. The accumulation of transit routes and services has constrained alternative uses of the landscape, forging relationships of interdependence between the isthmus and transportation clients. In order to understand the historical coproduction of physical geography, infrastructure, institutions, and social form, we need to return to the division of the seas.

The Isthmus in Geological Time

Panama emerged over a subduction zone where two oceanic plates—the Caribbean Plate and the Pacific Plate—met and formed a volcanic arc in a contiguous American tropical ocean. The deep-water connection between the present-day Pacific and Atlantic began to close fifteen to twenty million years ago. Magma rose up through plate fissures, creating a broad ridge beneath thousands of feet of water. By eleven million years ago, an archipelago with marine and coastal habitats appeared across what is now the southern half of Central America. Sediment runoff gradually filled in the spaces between the islands and, by the end of the Miocene about five million years ago, only three water corridors linked the oceans. These corridors—known today as the Atrato basin of Colombia, the San Carlos basin of Costa Rica and Nicaragua, and the Chagres basin of Panama—would become sites where human societies envisioned and built interoceanic transit routes millions of years later in an attempt to reverse geological processes.[7]

The Panamanian isthmus went from being a strait between the oceans to a land bridge between continents. The Chagres River basin was the site where the formerly contiguous ocean was divided and where it would be reunited millions of years later through the opening of the Panama Canal. The canal follows the old interoceanic corridor: a northwest-southeast depression about twenty miles wide. The separation of the oceans was an event of great climatic, geographic, and biological significance. The new Panamanian isthmus blocked the movement of marine life between the oceans and facilitated the movement of terrestrial organisms between North and South America, including the first humans to arrive from the north around ten thousand years ago. For the next four thousand years, a small nomadic human population probably lived in present-day Panama, moving across a large territory as they hunted, fished, and foraged. Agriculture arrived in Panama an estimated seven thousand years ago, changing how human groups interacted with the environment.[8] Indigenous people adopted swidden methods and grew corn, which originated in Mexico. Scholars debate the size of indigenous populations in the Americas and the degree to which they transformed the environment.[9] It seems clear that pre-Columbian landscapes around the Chagres River were deeply anthropogenic, but the environmental transformations associated with fire and agriculture would pale in comparison to changes linked to transportation.[10]

Spanish Colonizers Search for a Strait, Build a Road

When Columbus's ship first reached the coast of Panama in 1502, he claimed the isthmus for Spain. European colonists sought a maritime connection to the wealth of Asia, an orientation distinct from that of indigenous peoples, flora, and fauna traveling overland between North and South America. For the first three decades of Spanish colonial exploration in the Americas, they tried in vain to find a strait across the land that they called Veragua. According to legend, indigenous people told Columbus of a "narrow place" by a mighty river that led to the Indian Ocean.[11] He and others scoured the coast, but never found the mythical strait. When Vasco Núñez de Balboa—Spanish explorer, conquistador, and governor— saw the Pacific Ocean in Panama in 1513, he was reportedly the first European to do so.

In the absence of a natural strait between the oceans, the Spanish began actively reorganizing the Panamanian isthmus as part of a transportation artery between Europe and their colonial possessions in western South America. This began with a river and a road. Colonialism and imperialism have been bound up with many rivers—from the Mississippi in North America to the Congo in Africa—that have linked the coast with the interior. The Chagres River was different because it was not used to extract resources from Panamanian hinterlands, but to move wealth between oceans and continents. In 1514, one year after Balboa saw the Pacific, the governor of Castilla de Oro (Panama was then part of the Spanish colony of Peru) requested permission to import slaves to the isthmus to build a paved road across the headwaters of the Chagres River to facilitate the transportation of precious metals from Peruvian mines to Spain.[12] The Spanish established port cities—Panama on the Pacific and Nombre de Dios on the Atlantic—and used slave labor to build the Camino Real (royal road) between them. Even after the road was built, the Spanish continued to search for the mythical "hidden strait." In 1527, however, Captain Hernando de la Serna first surveyed the Chagres and reported that the river was navigable by ship thirty miles from its Atlantic mouth and by canoe and flat-bottomed boat at all points.[13] When it became clear that there was no strait, the river itself became a conduit for interoceanic transportation.

By the 1530s, two routes connected the oceans: overland on the Camino Real or a combination of land and water travel via the Chagres River and Camino de Cruces (road of crosses) (figures 5.1 and 5.2). The Camino Real was fifty miles long and roughly paved with stone blocks. A mule trip through the mountainous headwaters of the Chagres River took four days. Travelers described the journey as "18 leagues of misery and curses" because rain and traffic would turn the road into impassable mud and pirates, escaped slaves, and others often raided the passing mule trains.[14]

Panama's first transportation service infrastructure grew up along the colonial roads. By 1535, three roadside inns were in operation, offering lodging and food. For over ten miles, the Camino Real followed the Boquerón River, passing by a village named Boquerón, near where I conducted fieldwork almost five centuries later, and a mule station at Peluca—the current location of a Panama Canal's hydrographic station (chapter 2). The second Atlantic-to-Pacific option was to travel by water up the Chagres River to the town of Cruces (near the current community of Gamboa), located at a bend in the river, and then continue on to Panama City by foot on the Camino de Cruces, which was both more level and better surfaced than the Camino Real.[15] A transit boom began along the roads

Figure 5.2
The old Camino de Cruces trail, 1914. *Source:* National Archives at College Park, Maryland, RG 185-G, Box 5, Vol. 9.

and river in the early sixteenth century and lasted until the second half of the seventeenth century. Because no single social group controlled transportation networks, several groups benefited, including the urban commercial class, storehouse owners, and *transportistas*—the workers who physically moved foreign wealth by driving mule trains over the *caminos* and paddling boats up and down the Chagres River.[16] At that time, human work was as important as built infrastructure for facilitating movement across the isthmus.

The modification of the Chagres River for navigation—a precursor to canal construction—began around the time of the construction of the Camino Real, when the river was dredged to facilitate the passage of larger and deeper boats. It could run so shallow in the dry season that even small vessels had trouble traveling at points upstream. In 1534, the Spanish dredged the Chagres River channel, removing sunken trees and other obstructions, and King Charles requested that the governor explore possibilities for connecting this Atlantic-bound river to the Pacific.[17]

The seasons shaped travel via the early interoceanic land and water routes, foreshadowing issues around rainfall that still persist to this day. The Camino Real was dry and best to travel when the river was lowest. The river was high and easy to navigate when the road was muddy and impassable, so the land and water routes were complementary. In a watery environment, all of this—the construction, the maintenance, the transportation itself—required a great deal of labor. The governor imported black slaves to build and maintain the roads and work on barges and mule trains carrying goods between the oceans.[18]

Panama's boom went bust in the eighteenth century after pirates, attracted by the vast colonial wealth funneled across the isthmus, attacked the port cities of Panama and Portobelo. In response, the Spanish crown redirected colonial ship traffic to the more secure all-water Cape Horn route, so the rest of the eighteenth century and early nineteenth century brought little traffic to Panama.[19]

Panamanian Aspirations and Yankee Ambitions

Upon its independence from Spain in 1821, Panama became a territory of the short-lived Gran Colombia, which also included the present-day nations of Colombia, Ecuador, and Venezuela. By 1830, Gran Colombia had dissolved and Panama joined Nueva Granada, which was governed from Bogotá. Panama was free from colonial rule and, although the power structure of that era—namely, the hegemony of the merchant class—persisted, the colonial transportation infrastructure did not. The Panamanian elite of the early nineteenth century wanted to return the isthmus to its past commercial prominence, but they did not have the capital to repair its crumbling roads or undertake large construction projects like a canal or railroad themselves, so they pursued foreign capital. Overall, traffic across Panama was miniscule between 1821 and the 1840s.[20]

In the early nineteenth century, the US government was expanding its territorial holdings in North America and beyond. At that time, both the state and capitalists began to see a new role for the isthmus as an interoceanic route linking the eastern and western United States, rather than South America and Europe. Meanwhile, new technologies like the railroad and the steamship made the movement of people and goods across greater distances faster and more affordable. The British began steamship service to Panama in 1842 and, four years later, the United States annexed Oregon and took a greater interest in the isthmus, because overland transportation to the Pacific Northwest via stagecoach and horseback was slow and difficult. The governments of Nueva Granada and the United States signed a treaty in which the former guaranteed the latter the right of transit across Panama under the condition that the United States would guarantee Nueva Granada's sovereignty over said territory.[21] Gold was discovered in California at the end of 1848, which meant that the number of people seeking transportation to the Pacific exploded, practically overnight. During the first years of the rush that ensued, overland transportation or ship travel around Cape Horn took at least four months, but travel through Panama could be completed in as little as six weeks.[22] Thus began another construction and transportation boom in Panama.

The Panama Railroad and the California Gold Rush

The Panama Railroad Company was incorporated in New York in 1849. The company planned to build the world's first transcontinental railroad on the isthmus. Preparations were underway before news of gold in California reached the East Coast, but the discovery made the project more significant. By the time the railroad company broke ground in 1850, a new transportation boom—the first since the colonial era—was underway on the isthmus. The boom's beneficiaries included the urban merchant class of Spanish descent who hoped to revitalize the colonial transit economy and a larger population of *gente de color*—people of African, indigenous, and mixed descent—that controlled key parts of a transportation network little changed from previous centuries.[23]

When forty-niners bound for California arrived in Panama, their passage proceeded through several stages, each controlled by a different group of workers. First, boatmen carried them from ship to shore at the mouth of the Chagres (the harbor was too shallow for oceangoing steamships). Second, other boatmen would transport the travelers up the river in canoes or flat-bottomed boats to the town of Gatun (one day's travel) or as far as Cruces. Third, the forty-niners would proceed overland to Panama City, paying *cargadores* (porters) to carry their belongings. As before, navigating the Chagres River upstream to Cruces could be difficult during the dry season due to low water levels, even in a small boat.[24]

Figure 5.3
Boat trip on the Chagres River before the Panama Canal. *Source:* Personal collection of Doug Allen.

Many of the key challenges of railroad construction were not technical, but social. In the official history of the Panama Railroad, F. N. Otis wrote, "Instead of a secluded and rarely-visited region, where laborers and materials such as the country afforded were comparatively inexpensive, as was the case when the [railroad] contract was framed ... [Panama] was now swarming with emigrants from all parts of the globe *en route* for the land of gold. The conditions under which the contract was entered into were changed, the whole *morale* of the country had assumed an entirely different aspect."[25] Because of the numerous opportunities for people to make money working independently in the transport economy, the railroad company had difficulty recruiting and maintaining a labor force, so they imported workers from Colombia, Europe, the West Indies, and China.

The administrators of the Panama Railroad Company—a centralized organization assembled around a single line—conspired to undermine the established and decentralized network of boats and mule trains to control transit and drive "natives" into construction labor. In 1849, the company introduced small steamers on the lower Chagres able to carry a hundred passengers each, dwarfing the capacity of small local craft. The steamers charged lower prices, forcing independent boatmen out of business or to move further upstream. In this case, the hydrology of the river gave rise to aquatic spaces that could be monetized by different actors depending on their technologies and knowledge. Its seasonal ebb and flow protected the local industry upstream from steam power, because steamers could not travel in the shallowest areas. However, competition became fierce upstream, too, when the railroad company conspired with land transportation companies to drive local muleteers out of business.[26]

The construction of a railroad line between Aspinwall (now Colón) and Panama City began in 1850 and was completed in 1855 at a cost of eight million dollars—twice the duration and six times the expense estimated.[27] The city of Colón, which would become the Atlantic terminus of the Panama Canal, was constructed by draining the marshy Manzanillo Island, adding landfill, and linking it to the mainland with a causeway. The number of lives lost during railroad construction is unknown, but historians have estimated six thousand deaths.[28] The train carried an estimated four hundred thousand passengers during its first decade of operation.[29] By combining steamship travel with train travel across Panama,

Figure 5.4
Panama Railroad station at Bas Obispo, 1904. *Source:* National Archives at College Park, Maryland, RG 185-G, Box 4, Vol. 8.

it was possible to get from New York to California in twenty-three days, one-fifth the time of a journey around Cape Horn.[30]

The rise of the railroad transformed how the benefits of the transportation economy were distributed on the isthmus. Before the train, local people participated in the transportation economy by working in or between its loosely connected stages, but the railroad moved passengers rapidly from ocean to ocean, bypassing the inns, bars, and river towns in between. Moreover, the operation of the train required only a fraction of the many thousands of workers employed during construction.[31] Its conclusion left large numbers of people, immigrants and Panamanians alike, without work in the midst of a boom.

The economic decline of the train route was as precipitous as its ascent. The competition between interoceanic transportation routes was both regional (river vs. rail) and international (rail vs. rail). In 1869, the first transcontinental railroad line across the United States was completed and the advantages of the Panama Railroad for intercoastal US travelers were nullified. Company revenue fell from four million dollars in 1868 to one million in 1871.[32] Panamanian historian Alfredo Castillero Calvo concludes that the railroad boom and bust repeated a history familiar to Panamanians: foreign capitalists or governments in the terminus cities of Panama City and Colón absorbed most of the benefits of the boom. Yet the railroad also marked a different engagement with the global than mule trains and river boats because of the specific ways the technology articulated with society.

The railroad reorganized a decentralized transit economy around a single line and wage structure controlled by one company, hollowing out the rural economy. The tracks ran through rural areas along the Chagres River and the train paused at established stops along the way, but there was little impetus for passengers to spend money there. Profits were channeled to the steamship companies, the railroad company, and businesses in Panama City and Colón, where ship and train lines intersected. But, as it turned out, the railroad could also transport Panamanian products around the world.

Bananas along the Railroad Tracks

For centuries, Panama's transportation infrastructure has both shaped and been shaped by power asymmetries. Yet, despite the fact that interoceanic routes were not generally built for Panamanian use, they transformed the landscape in ways that created problems and provided opportunities for people on the isthmus. In 1866, Carl Franc, a German steward for Pacific Mail Steamships, entered an agreement to ship bananas from Colón to New York City.[33] Frederick Adams, historian and eulogist of the United Fruit Company, wrote of Franc and company, "They had preempted the only known spot in the American tropics where it seemed safe to raise and export bananas. The great stream of the world's commerce beat up against Colón. The Panama railroad was in operation, and the demands of international trade automatically decreed that peace and order should prevail in the territory adjacent to that natural pathway of commerce."[34]

Bananas flourished in the rich alluvial soils along the railroad, but their success was not natural. The fruit thrived due to a complex set of technical, ecological, and social conditions that were entangled with transportation routes. Yet Adams's framing elides the importance of the human communities that made banana exports possible. Among the region's economic advantages in the 1860s were thousands of unemployed former railroad laborers who might work in banana production. Some migrant laborers settled in the cities of Colón and Panama City, but many returned to agriculture and rural life in Panama.[35]

Figure 5.5
The Chagres River with the old village of Gatun on the far bank, 1907. *Source:* National Archives at College Park, Maryland, RG 185-G, Box 10, Vol. 19.

The village of Gatun—a day's travel upriver from the Atlantic during the colonial era and the first railroad stop in the late nineteenth century—was the hub of the rural banana trade, but North American travelers saw the community as a historical footnote. Historian F. N. Otis wrote, "The ancient native town of *Gatun*, which is composed of forty or fifty huts of cane and palm ... is worthy of mention as a point where, in the days by-gone, the bongo-loads of California travelers used to stop for refreshment on their way up the river."[36] The economy of the *Gatuneros* revolved around transportation, but became increasingly linked to the nascent banana trade after the completion of the railroad in 1855. Smallholders arrived by river one day each week from farms in the vicinity in long wooden canoes stacked high with bananas. From Gatun, bananas were shipped to the port in Colón and loaded on steamers bound for the eastern United States. Franc's Aspinwall Fruit Company flourished during the late 1860s and 1870s, a stopgap period between the decline of the railroad and beginning of the French canal project. At this point, infrastructure along the Chagres River could be mobilized to support the regional agricultural economy, but that arrangement would not last.

Figure 5.6
A smallholder banana market on the Chagres River. *Source:* Personal collection of Doug Allen.

The French Sea-Level Canal Project

Europeans had dreamed of building an interoceanic navigation canal across the Americas since the dawn of colonialism. Yet, despite the economic, political, and cultural attractions associated with this ambitious endeavor, nobody actually attempted it until late in the nineteenth century, after a series of large engineering successes boosted public confidence in the capacity of modern nations and firms to complete projects that had once seemed impossible. The Erie Canal, completed in 1825, linked New York to the Midwest via the Great Lakes and reorganized the economic geography of North America. The Suez Canal was opened in 1869 and brought Europe some six thousand miles closer to Asia by water, a boon for European manufacturers seeking to access new resources and markets.

Canals were conduits for economic profit, but they were also thought to carry social progress. They were arguably even more symbolically charged than other transportation infrastructure like roads and railroads in that they entailed the transformation of the earth itself to facilitate movement. The French effort to build a sea-level canal across Panama married geopolitical aspirations, economic motivations, and cultural beliefs in the Euro-American civilizing mission.

In 1879, the Congres International d'Etudes du Canal Interocéanique convened in Paris to discuss an isthmian canal project. Presiding over the meetings was Ferdinand de Lesseps, most famous for "building" the Suez Canal. Now, de Lesseps wanted to build a second major canal in Panama. He was a particular type of canal builder. He was not an engineer, like the West Point technocrats that controlled the Army Corps of Engineers. Nor was he a financier like those that built the Panama Railroad. He was a promoter, first and foremost, and was at his best conjuring collective imaginaries to sell projects.

Before the congress, Lucien Napoléon Bonaparte-Wyse and Armand Reclus, both young French naval officers and engineers, conducted surveys of potential canal routes in San Blas and Darién (in eastern Panama, near Colombia) and the Chagres River valley. Bonaparte-Wyse met with the Colombian president and signed an agreement that granted a ninety-nine-year concession to a group including de Lesseps to build a canal across Panama. The agreement stated that the Colombian government was to receive a portion of annual revenue not less than two hundred and fifty thousand dollars and that the canal route should be selected by an international congress of engineers. So, a year later, a body of 136 experts from France, its colonies, and 22 other countries assembled to discuss the Panama project. Bonaparte-Wyse, the naval engineer and surveyor, assembled data on seven potential routes for consideration at the congress. Four of the proposed routes crossed San Blas and the Darién and were expected to require some combination of locks or tunnels. Two routes—a lock canal and a sea-level canal—traced the route of the Panama Railroad near the Chagres. The final option was a lock-canal across Nicaragua.[37]

The congress followed a model that de Lesseps had used during the Suez Canal project. The delegates were divided into five committees focused on different aspects of the project and asked to assess the proposed routes and make a recommendation. The central drama unfolded in the technical committee. The US delegation—the largest after the French—supported a lock canal in Nicaragua. But de Lesseps was fixated on building a sea-level canal across Panama, like Suez. Supporters of the sea-level proposal argued that it would have many benefits over a lock canal design. For example, it could operate twenty-four hours per day and would not restrict the number of passing ships. But could it be built?

The engineers in attendance disagreed on the practicality of the sea-level canal in terms of cost, climate, and the volatility of the Chagres River, which would drop precipitously into the shipping channel and potentially threaten passing vessels. But, for de Lesseps, the lock design presented a different and more social problem: visibility. He did not believe that it would capture the public imagination to the same degree as excavating a ditch between the seas.[38]

After great debate, the technical committee recommended the sea-level route, a choice overwhelmingly approved by the congress as a whole. Ferdinand de Lesseps agreed to take charge of a project that was estimated to cost two hundred and fifty million dollars and take twelve years to complete, but skeptics of the plan remained.[39] Baron Godin de Lepinay, a senior French civil engineer who recommended a lock canal with a large summit lake (basically the future US design), was among those who predicted imminent failure. He reminded delegates that environmental differences are of great consequence in engineering projects:

They want this canal to be made after the mode of the Suez Canal, that is to say, without locks—and yet its natural conditions are so very different. In Suez there is no water, the soil is soft, the country is almost on the level of the sea; in spite of the heat, the climate is perfectly healthy. In tropical America there is too much water, the rocks are exceedingly hard, the soil is very hilly, and the climate is deadly. The country is literally poisoned. Now to act thus after the same fashion under such different circumstances is to try to do violence to nature instead of aiding it, which is the principal purpose of the art of engineering.[40]

Unlike Panama, the Suez site was also near large population centers from which laborers could be imported. What these distinctions pointed to was that everything learned by building a canal across Egypt had to be unlearned in Panama.[41]

The French canal company began canal construction in 1881. It was a private enterprise funded by wealthy shareholders and thousands of French bondholders, who supported the project in hopes of financial gain, but also to participate in a "civilizing" project. Excavation proceeded haltingly and at great expense over the next seven years, as the company experimented with different contractors, large and small. The environmental and human problems that the naysayers of the Panama route had predicted—water, soil, disease, labor—all came to pass. The Chagres was prone to erratic floods during the rainy season. Moreover, due to the geological history of the isthmus, the route crossed six major faults and seventeen types of rock—including clay, gravel, sand, limestone, and coral—meaning that excavation posed different challenges along the line.[42]

Labor was an intractable problem, just as it had been for the railroad builders thirty years before. "Native" workers were reportedly uninterested in construction wage labor because of the many opportunities for work available in the decentralized transportation economy. Therefore, the French recruited a work force in the West Indies. The work force grew from 2,000 in 1881 to 20,000 in 1884, some 60 percent of whom were West Indians from Jamaica and Barbados. Many West Indian laborers descended from slaves emancipated in the 1830s, who subsequently went to work in smallholder or plantation agriculture for little pay. So the opportunity to dig the canal in Panama—with free passage and a salary many times higher than those at home—was attractive.[43] From a Euro-American managerial perspective, West Indians of African descent were thought to be adapted to Panama's tropical climate and to be subservient because of their history with the British colonial system. In all, the French canal company employed fifty thousand West Indians in construction.[44]

When the French canal company went bankrupt in 1889, it had spent 287 million dollars on a partially excavated ditch—nearly 40 million dollars over budget. The year before it folded, the company had abandoned its ambitious plans for a sea-level canal in favor of a lock design that demanded less excavation, but was unable to attract more investors before funds ran out.[45] Charges of financial mismanagement and corruption circled the company and those involved. Even the once unimpeachable de Lesseps was found guilty of misappropriation of funds. On the isthmus, the limits of the French canal builders' engineering, medical, and environmental knowledge plagued the project. Work had begun with inadequate understanding of geology and hydrology, a problem that could be linked to de Lesseps's gift for inspiring confidence—arguably too much—and limited engineering knowledge. Moreover, yellow fever and malaria ravaged the labor force, killing an estimated twenty thousand people.[46]

How the French Canal Became American

As the nineteenth century came to a close, the US government's imperial ambitions were growing. With the incorporation of the states of California, Oregon, and Washington, the government reached the edge of North America and expanded overseas by establishing military and commercial outposts in Puerto Rico, Cuba, Hawaii, the Philippines, and Guam. The government had been interested in an interoceanic canal in Central America since the mid-nineteenth century, a desire that was resurgent on the crest of the wave of expansionism.

The French project was moribund. After its bankruptcy in 1889, the company's extensive assets were transferred to a new entity, the Compagnie Nouvelle du Canal de Panama (New French Canal Company), which tried to sell them. However, no private company could afford to take on such a massive project and no government would defy the increasingly powerful and territorial United States. Therefore, the US government was the only potential buyer by default, but legislation had already been introduced in Congress to build a canal in Nicaragua. Legislators saw a Nicaragua canal project as a clean slate, while Panama's reputation was tainted by years of financial and technical problems.[47]

A complex chain of events that linked political intrigue, payoffs, and propaganda campaigns in Washington, D.C. led the US government to pursue a canal in Panama rather than Nicaragua. In 1901, the president, William McKinley, was assassinated and the vice president, Theodore Roosevelt, replaced him in office. Advocates for the Panama route convinced Roosevelt that it was preferable to Nicaragua from an engineering standpoint. Panama's main advantages were not natural, but historical: better ports, an established railway, and the infrastructural legacies of a decade of construction work. Panama was a worn pathway. Nicaragua, by contrast, was a less-known and less-developed route.[48]

Figure 5.7
Abandoned French dredge in channel between Pedro Miguel and Miraflores, 1913. *Source:* National Archives at College Park, Maryland, RG 185-G, Box 7, Vol. 14.

The Colombian concession to the French canal company restricted its sale or transfer to other private companies; national governments were off-limits. The US government approached Colombia and offered ten million dollars and an annual payment of two hundred and fifty thousand dollars for a hundred-year renewable lease, but the legislature in Bogotá rejected the proposal.[49] This left the Roosevelt administration with an undesirable set of options. They could continue to negotiate with Bogotá, pursue the Nicaragua route, or begin construction on the isthmus without a Colombian concession. For the US government to build a canal in Panama without Colombian approval would amount to a rejection of the terms of the 1846 agreement between the nations that guaranteed the United States and its citizens free passage across the isthmus in exchange for a guarantee of the rights of Colombia to administer that territory. The Roosevelt administration sidestepped that obligation by conspiring with a group of Panamanian nationalists seeking independence from Colombia. In fact, Panamanian groups had attempted secession four times in the nineteenth century.[50] But Colombia's rejection of the US canal proposal provided a unique opportunity to make those dreams real.[51]

In 1903, Panama declared independence from Colombia. Hours later, US gunboats crowded its harbors and troops landed on the isthmus. The news was met with anger in Colombia. After the revolution was complete, the Roosevelt administration argued, dubiously, that it had not broken the terms of the 1846 treaty, even though it had purposefully undermined Colombian sovereignty. The government's shaky legal justification was that the treaty was "a covenant with the land"—the territory itself—rather than Colombia. The United States also invoked the global dimensions of the unbuilt infrastructure, claiming that, by refusing its terms, Colombia was blocking a "future highway of civilization."[52] The US government began negotiations with the newly sovereign Panama to build a canal. Much of the drama that ensued unfolded in Washington, D.C., New York, Paris, and Bogotá, centers of power far from construction works.

The Afterlife of Infrastructure Projects

From an engineering and financial perspective, the Panama Railroad was a success and the French canal project a failure. On the ground, however, both projects were the same in at least one critical way: they came to an end and left thousands of laborers unemployed. Like the railroad before it, the French canal project left some thirteen thousand West Indians stranded in Panama without work.[53] Large numbers of former laborers from

Table 5.1
Estimated Population of the Chagres River Watershed, 1790–1896

Year	Population	Density (people/km)
1790	1,500	0.5
1851	2,000	0.6
1896	20,000	6.0

Source: Jaén Suárez, Hombres y Ecología en Panamá, 137.

other countries remained on the isthmus. By 1896, the area adjacent to the railroad and river—called "the line"—was inhabited by twenty thousand people, largely West Indians and Colombians of African descent, and smaller numbers of Europeans and Asian migrants (table 5.1).[54]

Through the upheavals of the nineteenth century, the village of Gatun—a cluster of small wooden buildings with zinc roofs shaded by fruit trees and coconut palms—remained a center of local life on the river, even if it was overshadowed by the port cities. It was home to railroad workers in the 1850s, headquarters of the French canal project in the 1880s, and a key node in the banana economy that served as a stopgap between construction projects. Banana export production in Gatun nearly ceased in 1881 when the French canal company offered wages high enough to motivate farmers to abandon their fields from the Chagres River to Costa Rica.[55] But many returned to bananas after the project failed and workers were laid off en masse. As late as 1904, Panamanian officials reported that Gatun participated heavily in the banana trade: seven to nine railroad carloads of bananas were shipped from the town weekly.[56] That year, however, construction took center stage again as the US canal project began. Banana plants would not be cultivated in the valley in large numbers for two decades. When they were, it was under very different circumstances (chapter 8).

Before the US canal project began, the boundary between the transport and agrarian economies was permeable. For centuries, ethnically and culturally heterogeneous populations that Euro-Americans glossed as "natives" moved periodically between wage labor and farming as economic conditions changed. During the new project, however, infrastructure construction radically transformed former agricultural landscapes and the US government pursued a territorial politics that emphasized the enclosure of the transit zone.

Conclusion: The Sedimentation of Infrastructure

Transportation has dominated the economy around the Chagres River for long stretches since the sixteenth century, when the isthmus first became a passage point for colonial wealth extracted from South America and bound for Spain.[57] The Panamanian historian Alfredo Castillero Calvo characterizes the long-running relationship between Panama and the global economy as *transitismo* (transitism). He argues that Panama is different from most countries across the global periphery and Latin America because, for centuries, its economy has been organized around transportation services—*la vocación geográfico* (geographic vocation)—rather than resource extraction or monoculture agriculture. Thus, Panama's past and present must be understood in terms of specific relationships of dependency with Spain and, later, the United States formed around transit.[58]

For Castillero Calvo, the "vocation" of the isthmus as a transit zone within the political-economic structure of the modern world system has had a number of significant consequences. Capital and population have been concentrated in the transit zone and power in the hands of a commercial class rather than the landed elite who historically controlled many Latin American countries. Moreover, economic alternatives like agriculture and industry have been stifled, leaving the isthmus dangerously dependent on unstable external markets for food and manufactured goods and producing disequilibrium among its various regions. *Transitismo* provides a wide-angle perspective on economic patterns over centuries sensitive to the specificities of Panamanian history, but like other Marxian global theories, it assigns the global economy an internal logic and overarching power that obscures the social and material construction of "global" infrastructure in Panama. In the chapters that follow, then, I nuance, extend, and complicate the *transitismo* argument through attention to the infrastructural work that has transformed the isthmian landscape in expected and unexpected ways.

6 Canal Construction and the Politics of Water

Figure 6.1
Unfinished Pedro Miguel Locks in a rainy season flood, December 1910. *Source:* National Archives at College Park, Maryland, RG 185-G, Box 3, Vol. 6.

Before the canal, there was the river. The Chagres was the central artery of a region veined with roads and railroads funded by colonial and imperial powers and built and maintained by migrant laborers. The river also had a life of its own. It flooded in the rainy season and ran shallow during the dry season, thereby exerting agency over travelers and the local boat pilots and mule train drivers that facilitated their journeys. The construction of the Panama Canal during the first two decades of the twentieth century was, in large part, an effort to discipline the river by controlling and redirecting its flow to serve year-round navigation. While the excavation of the canal has been well documented, little attention has been paid to the US government's simultaneous efforts to create a transportation environment by transforming physical and human geography across the entire river basin. During the 1904 to 1914 construction era, the government resettled thousands of people from rural areas of the new Canal Zone to establish waterscapes for navigation and legible landscapes to facilitate governance.

The technologies assembled to reorganize the Chagres River as a water source for the lock canal fixed the volume of water per ship transit at fifty-two million gallons. However, water was not managed exclusively for navigation—that is, as a presocial resource that became an object of human politics. Liquid was materially imbricated in the infrastructures through which governmental practices unfolded—some territorial, others biopolitical.[1] Water infrastructure bridged engineered technologies (locks and dams), environmental expertise (watershed surveys, hydrographic data collection, and cartography), and political technologies (enumeration, resettlement, and monitoring), linking imperial land and resource enclosures to intimate efforts to control human bodies and the ecology of the household.

Empire is, at its root, a territorial project. Its architects establish outposts far from home from which to project power. Once a suitable site is located, the next step involves the expropriation of land, often by coerced consent.[2] The imperial state then establishes and extends its authority through governmental techniques aimed at both large human populations and more intimate spaces (from military interventions and censuses to the formalization of racial taxonomies) and the territorial politics of landscape transformation (environmental management, sanitation, public works, and engineering projects). As we will see, imperial technologies and territoriality were inextricable in the space of the US Canal Zone.

In the Panama Canal Treaty of 1903, as explained in chapter 1, Panama granted the US government the "use, occupation, and control" in perpetuity of a ten-mile-wide strip of territory bisecting the narrowest part of the isthmus. The political boundaries of that space, the Canal Zone, were not fixed. In addition to near-sovereign powers within the original strip, the treaty gave the United States the authority to expropriate additional lands

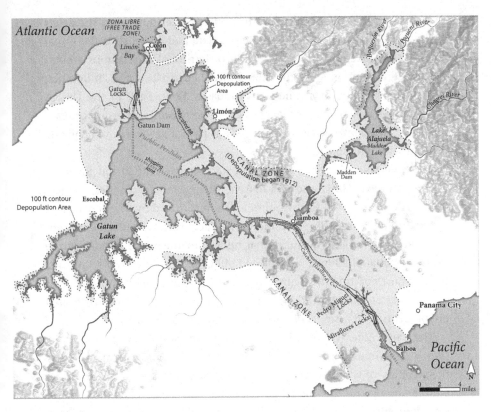

Figure 6.2
The Canal Zone and canal's water management system in the early twentieth century. Image by Tim Stallman, used with permission.

and waters as necessary for the "construction, maintenance, operation, sanitation, and protection" of the canal.[3]

Far more than an engineering project, the Canal Zone was defined as an infrastructural space with a completely open-ended definition of what was actually "necessary" for canal purposes. The US government effectively had a lien on additional lands and waters in the Republic of Panama beyond the Zone's initial boundaries that might justifiably *become* necessary for transportation purposes in the future.[4] This led to decades of US territorial expansion and related conflict on the isthmus.[5] The search for and management of fresh water for navigation, formalized in a 1903 treaty clause that granted the US state expansive rights to Panamanian rivers, streams, and lakes, would be central to US expansionism and the reorganization of this space.

With the canal treaty in place, the United States purchased the assets of the French project for forty million dollars and established the Canal Zone in the wake of a rapid succession of territorial acquisitions around the planet, including: Alaska in 1867; Hawaii, Puerto Rico, the Philippines, and Guam in 1898; and American Samoa in 1899. Back in the United States, critics called overseas expansion imperialism, while supporters— especially President Roosevelt—framed it as part of an American civilizing mission with widespread benefits.[6]

The overseas territories of the United States were classified and governed in different ways. Alaska and Hawaii were "incorporated territories," subject to provisions of the US Constitution and administered by the Department of the Interior. The Philippines and Puerto Rico were "organized but unincorporated territories," meaning that only fundamental provisions of the Constitution applied (e.g., a truncated bill of rights). They were governed by the War Department. Finally, the Canal Zone, like Guam and American Samoa, was classified as an "unorganized possession," given a truncated bill of rights, and assigned a government by executive order.

The Canal Zone was initially governed by the Isthmian Canal Commission, a seven-member board based in Washington, D.C. that answered to the secretary of war. However, slow communication between Washington and Panama made this model cumbersome and led to construction and supply problems on the isthmus. The commission was then reconfigured as a three-member board—a chairman, a chief engineer, and a governor— to be based in Panama. In 1907, Major George Goethals, formerly of the Army Corps of Engineers, was designated chief engineer and chairman of the commission. He was the third chief engineer in four years, following John Wallace (1904–1905) and John Stevens (1905–1907). In 1908, the government stripped the commission of its authority and replaced it with an autocratic government headed by Goethals that controlled every aspect of life in the Zone, including housing, food, transportation, sanitation, health care, and recreation.[7]

From Soil Excavation to Water Management

"Let the dirt fly" was a popular refrain during the construction of the Panama Canal. However, the opening of the waterway also entailed the flooding of hundreds of square miles of Panamanian land to store water for shipping and the forced resettlement of dozens of communities. Unlike soil excavation, then, a focus on water management reminds us that infrastructure is entangled with landscapes and livelihoods that blur system boundaries. Thus, to attend to the establishment of the Panama Canal's water management infrastructure draws our attention to the unfinished quality of the historical technical systems that we live with.

The central site of Panama Canal excavation was the nine-mile-long Culebra Cut—later renamed Gaillard Cut—carved through the isthmus's continental divide to divert some of the water running through the Atlantic-bound Chagres River south and into the Pacific Ocean. Most historians of the construction effort have focused on the Culebra Cut because of the drama inherent in a monumental struggle involving throngs of laborers (up to six thousand working at one time), hulking machinery and complex technical networks, and earth that seemed to actively resist the opening of the waterway by sliding into the cut as it was excavated.[8]

Figure 6.3
Tourists flocked to Panama to see canal excavation work. *Source*: Personal collection of Doug Allen.

The Culebra Cut was not a monolithic excavation effort, but a multitude of projects organized in terrace-like levels along the walls of a manmade chasm. Each level had its own assemblage of railroad tracks, steam shovels, loaders, track shifters, and labor gangs in blue shirts and khaki pants.[9] Together, laborers and machines drilled, blasted, shoveled, excavated, and transported soil. Up to one hundred and sixty trains per day moved over a web of tracks between the cut and regional dump sites.[10] For workers—particularly the black West Indian laborers that dominated the ranks of the "silver roll" at the bottom of the canal's racialized and nationalized labor hierarchy—the cut was an extremely uncomfortable and potentially fatal work site. During the dry season, temperatures reached one hundred and twenty degrees Fahrenheit with little shade available. During the rainy season, torrential downpours turned soil into mud and laborers toiled in wet clothes day after day, conditions that led many to die of pneumonia.[11]

The cut roared with machinery and reverberated with explosions. Sixty-one million pounds of dynamite were used during construction—more explosive energy than the United States had deployed in all of its previous wars combined.[12] And over half of the workforce was involved in dynamite work, which left many laborers maimed or dead. These and other industrial accidents—like errant trains that ran over workers—punctuated the brutal monotony of construction. A visiting scientist described the scene in this way:

I am more afraid of the blasting than anything else. They become a little reckless I think. Dynamite is used by the ton. One day I was on a labor train and some shots were fired almost alongside and rocks flew over us and out into the lake a hundred yards away. They blow whistles—usually—as a warning, but you do not always know where the shots are so it is a good plan to get under something as soon as you can. One day I was partway across the Gatun dam and the engines began to whistle and I saw men running in various directions. I was not very near any shelter and did not know where the shots were. They went off uncomfortably near.[13]

Whatever romance was associated with the old dream of digging a sea-level "ditch" across Panama seemed to affix itself to its closest approximation, because the Culebra Cut attracted a level of public interest that dwarfed other canal works. Historian David McCullough writes,

It was the great focus of attention, regardless of whatever else was happening at Panama. The building of Gatun Dam or the construction of the locks, projects of colossal scale and expense, were always of secondary interest so long as the battle raged in the nine-mile stretch between Bas Obispo and Pedro Miguel ... in the dry season, the tourists came by the hundreds, by the thousands as time went on, to stand and watch from grassy vantage points hundreds of feet above it all. ... A spellbound public read of cracks opening in the ground, of heart-breaking landslides, of the bottom of the canal mysteriously rising. Whole mountainsides were being brought down with thunderous blasts of dynamite.[14]

Fifteen thousand tourists arrived in 1911 and twenty thousand in 1912 (figure 6.3).[15] The drama of men and machines struggling with the earth captured the imagination of tourists and those reading accounts at home.

Earth moving was the ultimate index of cultural progress or "measure of man" at the beginning of the twentieth century.[16] Thus, the completion of the canal was held up as a symbol of the triumph of modern man over the worldly constraints of tropical nature, climate, disease, and space—even geology itself—and excavation was central to that narrative. But water, not soil, was the element at the intersection of American technical initiatives to forge interoceanic connection across Panama and state efforts to produce a legible, governable space around the waterway.

The Lock Canal and Its Water Use Implications

In 1906, the US Congress voted to fund the construction of a lock canal. The American canal would, for the most part, follow the line of the partially excavated channel that the French canal company dug in the 1880s, but there was one significant difference. Before bankruptcy and the deaths of tens of thousands of laborers, the French company had planned to excavate a sea-level canal across the isthmus: a saltwater channel that, if completed, would have allowed ships to travel unimpeded between the oceans. By contrast, the American lock design would move ship traffic over a freshwater staircase of six locks (three up and three down).[17]

The lock decision was not inevitable. During the first years of construction (1904–1906), engineers debated the merits of sea-level versus lock canals, as they had in the nineteenth century. In 1905, President Roosevelt assembled a body of experts, the Board of Consulting Engineers for the Panama Canal, and charged them with recommending a design. The board members consulted existing surveys and reports on Panama and traveled to the isthmus to study the terrain firsthand. In the end, they were unable to reach a consensus and drafted two reports. The majority report advocated a second attempt to dig a sea-level canal and the minority report called for the construction of a lock canal. The sea-level project was estimated to cost 247 million dollars and take 12 to 13 years, while the lock plan was projected to cost 139 million dollars and require 9 years.[18]

Technical considerations, construction timetables, and budgetary concerns shaped canal design, but so did the dimensions of oceangoing ships at that time and projected into the future, climatic contingencies in Panama, and US cultural nationalism.[19] The members of the Board of Consulting Engineers who supported the sea-level design argued that, despite the French failure in the 1880s, an unobstructed waterway like the profitable Suez Canal could be built in Panama due to engineering advances since then and, moreover, that national pride compelled the United States, as Board Chairman Davis put it, to "treat this matter not in a provisional way but in a final masterly way."[20] Public belief in engineers' capacity to reorganize the planet to serve humanity seemed boundless.

The bright, if unrealized, future of canal building was trumpeted with breathless enthusiasm in venues like the *Chicago News* in 1908:

Canals more wonderful than those of Panama and Suez are already in prospect. The early summer is to see the commencement of an inland waterway that will dwarf them both in comparative significance. This is the long-projected Baltic and Black Sea canal, which will intersect Russia from north to south, a distance of one thousand miles, and the total estimated cost of which is put officially at five hundred million dollars. ... Another wonderful canal scheme which is being enthusiastically taken up in Italy contemplates nothing less than the joining by this means of Genoa and Lake Constance. To do this it will be necessary, of course, to cross not only the Apenine Mountains, but also the Alps. This it is proposed to do by means of a new invention of locks, involving the construction of a series of inclined tubular water lifts. It may yet be possible to travel by steamer over the loftiest mountain range in Europe.[21]

Panama's central mountain range, the Cordillera Central, was a humbler ascent than the Alps, but proponents of the isthmian lock canal also wanted to lift ships over mountains. The engineers modeled their design on the Soo Locks, two parallel flights of locks that allow ships to travel between Lake Huron and Lake Superior. They argued that the lock design was preferable to the sea-level canal in many ways: it would be safer for passing ships in Chagres River flood conditions, reduce the impact of landslides on transit, provide easier passage for large vessels, cost less to maintain, and be easier to enlarge and defend. John Stevens, the chief engineer of the canal commission, presciently concluded, "There is one valid argument, and only one, which can be brought against the canal with locks, and that is the difficulty of fixing the dimensions of the lock chambers to provide for the possible enlarged vessels of the future." [22]

Figure 6.4
Miraflores Locks under construction, January 1912. *Source:* National Archives at College Park, Maryland, RG 185-G, Box 3, Vol. 5.

After lengthy and charged deliberations in Washington, D.C., Congress approved a canal design organized around a large summit lake at eighty-five feet above sea level with flights of locks on either side. Thus, from the Atlantic Ocean, a ship in transit would ascend from sea level up to the lake by passing through three locks. After being released from the third lock, the ship would travel about twenty miles over the artificial lake—the flooded valley of the Chagres River—before entering the Culebra Cut and then descending to sea level through three more locks. After clearing the final lock, the ship would be released into the Pacific Ocean.

Locks, Shipping Standards, and Water Supply

The canal design fixed the dimensions of the lock chambers, a decision that has shaped both global transportation infrastructure (via ship design and waterway dredging) and Panamanian political ecology (through reservoir flooding and land use), as explained in preceding chapters.

Years before the design was finalized, the US Congress had stipulated in the Spooner Act of 1902 that any future canal afford passage to the largest ships in existence at the time of construction and "such as may be reasonably anticipated." But what time period was reasonable? Decades? Centuries? In the Board of Consulting Engineers minority report, the advocates of the lock canal argued for locks big enough to provide "reasonable but not excessive allowances for further developments," arguing that "no one would expect to provide in any commercial or military construction for needs at the end of the present century. ... If the locks are larger than necessary, they will not only cost more but will require a larger water supply and will not be quite so convenient to operate." [23] After negotiation between engineers, the government, and the military (particularly the Navy, which was concerned about the passage of battleships), the locks were set at 965 feet usable length, 106 feet usable width, and 39.5 feet usable draft. These dimensions created the Panamax shipping standard.

Figure 6.5
Water enters the Miraflores Locks for the first time, October 1913. *Source:* National Archives at College Park, Maryland, RG 185-G, Box 3, Vol. 5.

Ship size was not a problem when the locks were designed: 95 percent of oceangoing vessels at that time were under 600 feet long.[24] The largest commercial ships under construction in 1906 measured 800 feet long by 88 feet wide by 38 feet maximum draft, well within the proposed lock dimensions.[25] Yet, in 1908, before the locks were even built, the editors of Panama's *Star and Herald* were anticipating their obsolescence, writing, "what a discreditable and calamitous thing it would be to have at Panama a canal through which our best ships could not pass. ... The true policy is to remember that we are building this canal not for the present nor for the next ten years, but for all time, and that therefore we should build, as far as possible in a way which will meet all future needs."[26]

At first, the dimensions of ships shaped lock design but, as maritime transportation technologies changed and international trade increased, the Panamax standard—the maximum ship dimensions able to pass through the locks—shaped ship and port design at networked sites far from Panama.[27] Panama Canal engineers designed a water management network to meet the needs of contemporary and projected future shipping. In so doing, they also fixed a hydrological standard with profound implications for regional political ecology. After all, the locks were not simply part of a technical network, but a heterogeneous infrastructure that exceeded and reworked categories like nature, society, and technology. Ships traveled through harbors, locks, and shipping channels, all dredged deep enough to accommodate their underwater drafts. And, of course, the canal used rainwater that flowed through rivers and was stored in artificial reservoirs. The water supply demanded by this design—fifty-two million gallons per passage—precipitated the technological reorganization of the Chagres River and its tributaries to monitor, regulate, and consistently deliver enormous amounts of fresh water throughout the rainy and dry seasons.[28]

Seasonal Water Flow and Abbot's "Vital Question"

Given the lock design, US engineers needed to understand regional hydrology better than their predecessors. For the French—who sought to dig a canal full of ocean water—knowledge of the water volume flowing through surrounding rivers was important insomuch as floods presented a threat to navigation, but not in terms of available water supply. They built only three rainfall stations and three gauging stations along the length of the river for measuring water flow.[29] No instrumental survey of the Chagres drainage basin was conducted, so its area and runoff were only generally known. For a lock canal, by contrast, knowledge of the volume, speed, and variability of discharge, as well as watershed area, was critical.

"The vital question," wrote Henry Abbot, a hydrologist and retired Army Corps of Engineers brigadier general, in 1905, "was to determine whether the Chagres will supply all the needs of the Canal in seasons of low water. Any reasonable doubt here would be fatal to the project of a canal with locks."[30] Abbot was a leading expert on isthmian canal engineering debates, having served on both the US Board of Consulting Engineers and its French forerunner, the Comité Technique.

Abbot's "vital question" concerning future water supply seemed ludicrous to some foreign observers struck by Panama's torrential rainfall. For example, the US authors of a book entitled *The Story of Panama: The New Route to India* published during the canal construction era wrote, "The November visitor to the Zone who has seen the floods of the Chagres carrying before them trees, houses and bridges, submerging steam shovels, destroying miles of railroad, will never question the adequacy of the water supply."[31] Yet, as Abbot emphasized, precipitation is seasonal on the isthmus, which generally has a nine-month rainy season (April–December) and a three-month dry season (January–March).

Engineers emphasized fluctuations in rainfall and streamflow in lock canal designs because shipping required fresh water year-round. Shallow water in the Chagres River during the dry season had limited the upriver range of small boats for centuries; ships transiting the canal would be much larger. To ensure that the locks worked all year, an extensive technical system was assembled across the Chagres River basin. Its centerpiece would be Gatun Lake. Engineers envisioned the lake as a massive summit-level reservoir to collect water during the rainy season in order to operate the canal through the dry season. The reservoir's secondary purpose was to "tame" the volatile river by absorbing periodic floodwaters that might threaten passing ships.

US engineers only had the piecemeal watershed data collected by state- and privately-funded surveys that dated to the mid-nineteenth century.[32] They needed to know the Chagres River basin's area to estimate the water volume flowing into the lake in order to manage the water supply. In 1905, the Isthmian Canal Commission created the Bureau of Meteorological and River Hydraulics—the institution that later hired Frank Robinson (chapter 4)—to collect and analyze atmospheric, surface, and subsurface water supply data related to navigation. They consequently played an important role in embedding water in the technopolitical formations through which US state territoriality unfolded in Panama.

Mapping the Chagres River Basin

When William Silbert, head of the Department of Locks and Dam Construction wrote Chief Engineer Goethals to request funds for a hydrological survey of the Chagres River watershed in 1908, engineers were using a rough "sketch map" of the basin (1:300,000 scale) from the 1899–1901 Report of the Isthmian Canal Commission.[33] In 1908, the commission conducted a survey to more accurately determine the drainage area of the Chagres River—its watershed—in order to better estimate the volume of water that would flow into the future canal system. The Chagres River's headwaters were located beyond the Canal Zone in the Republic of Panama, so the assistant engineer in charge of the survey secured the necessary letters of introduction to Panamanian politicians in these rural localities and the required equipment: Y-levels, lining and stadia rods, compasses, hand levels, barometers, and field glasses. The first party entered the field in November 1908. This was, incidentally, the rainiest month of the year that soil excavation peaked in the Culebra Cut. As thousands toiled in the cut, three engineers and fifteen laborers pulled their canoes ashore along the Chilibre River and made camp.[34]

By the end of 1908, four parties of this size were surveying the watershed. Although surveying may not have been as dangerous as working on a dynamite crew in the Culebra Cut, the parties faced numerous environmental, social, and technical challenges. Flooding rivers made boat travel— the only mode of transit in the roadless area—very difficult. The boat operators and laborers hired to cut trails through the forest to lay survey lines quit unexpectedly or left for New Year's celebrations and never returned. Engineers became sick. Critical equipment broke and was sent away for repair. Days, even weeks, were lost due to such difficulties.

The field parties ultimately completed a traverse survey of the Chagres River proper up to a point about five miles below the watershed's ridgeline.

It was the most complete survey of the Chagres River watershed to date, building upon partial surveys conducted in 1864 by the Colombian government, 1875 by the US Navy, and between 1904 and 1907 by the Isthmian Canal Commission. In 1908, the surveyors located some basic points on the ridgeline and omitted some smaller rivers altogether. The new survey map was at a 1:150,000 scale—still rough, but it had twice the detail of the previous version. The Chagres River watershed survey established a body of hydrographic knowledge and cartographic representation that was central to the canal's development as a hydropolitical project.

Water Supply and the Territory of the Canal Zone

Water was not simply a natural resource. It was the locus of a number of US projects in Panama, ranging from the construction of dams, locks, and urban infrastructure to the expansion of US territory and the implementation of expansive public health and sanitation measures. Liquid presented different problems and opportunities for various canal institutions depending on their mandates, but, across the board, engineers, planners, and administrators shared a desire to transform "bad" water (volatile rivers, stagnant swamps, open cisterns where mosquitoes bred) into "good" managed water (orderly rivers, storage reservoirs, and municipal distribution systems). The technopolitics of water management emerged through the interwoven practices of engineering and governance.

Although the impetus for the 1908 Chagres River watershed survey was to collect hydrological data for engineering and water management, the cartographic knowledge produced was also a valuable tool for political interventions, particularly territorial expansion and the control of rural populations. This is why Chief Engineer Goethals, anticipating the expropriation of Panamanian land in the area, instructed the survey teams to expand data collection twenty feet above and below the surface level of the future lake.[35]

Gatun Lake was to be created by damming the Chagres River at a point upstream from its Atlantic mouth and flooding the surrounding river valley. This summit-level reservoir would serve the canal by storing river water to operate its locks and making up twenty miles of the forty-eight-mile-long shipping channel. The flooding of the lake transformed landscapes and livelihoods along the Chagres. US politicians invoked treaty rights to expropriate the area to be flooded and, in 1912, the Panamanian state recognized the right of the United States to administer an additional seventy square miles beyond the original Canal Zone for water management. Panama ceded land within the future footprint of the lake—to be maintained at eighty-five feet above sea level—and a buffer zone up to the hundred-foot contour, which became the new political boundary between the Zone and Panama (figure 6.2). The expropriation of land and water for Gatun Lake represented the expansion of US jurisdiction in Panama and affirmed how water management technologies and governmental techniques coalesced around topographical contours defined through survey maps. The lock design formatted the canal administration's approach to water management, which precipitated the flooding of Gatun Lake to a specific level, which, in turn, altered political boundaries. When the 1.5-mile-long earthen Gatun Dam was completed in 1911, the flow of the Chagres to the Atlantic was interrupted and the lake began to rise.

Table 6.1
Hydrology and Level of Gatun Lake by Year (1909–1915)

Year	Lake level (feet above sea level)	Rainfall inches
1909	3.63	162.42
1910	13.07	149.66
1911	15.15	98.41
1912	31.24	102.83
1913	57.87	102.40
1914	85.26	100.54
1915	86.17	118.17

Source: Henry Abbot, "Hydrology of the Isthmus of Panama," Proceedings of the National Academy of Sciences of the United States of America, June 17, 1917, 45.

Infrastructure, Settlement, and Land Tenure

Water infrastructure was shaped by political forces and then began to exert its own force. Dams and locks did not establish a transportation environment themselves, but the pursuit of a year-round water supply became the technical and political rationale for a broader state effort to remake an entire region in the image of navigation. Governance was a different kind of infrastructural work than engineering, but it was also part of reformatting regional landscapes to serve extralocal communities.

The US government had the power to expropriate land for "canal purposes," but, before 1912, they adopted a liberal policy toward rural settlement in the Canal Zone. President Roosevelt instructed the secretary of war in 1904, "The inhabitants of the Isthmian Canal Zone are entitled to security in their persons, property and religion, and in all their private rights and relations. ... The people should be disturbed as little as possible in their customs and avocations that are in harmony with principles of well ordered and decent living."[36] This attitude would change as the construction of the canal proceeded and the US government and canal administration sought to remove inhabitants, regardless of how they lived.

In 1908, the US government controlled two-thirds of the land in the Canal Zone. They had purchased the land owned by the New French Canal Company (52 square miles) and Panama Railroad Company (68 square miles) outright and, through the treaty, secured a lien on Panamanian public lands (189 square miles), but they did not own the significant land area under private title (136 square miles).[37] The original canal treaty stated that the United States was required to compensate those with private land titles for damages associated with the construction, maintenance, operation, sanitation, and protection of the canal.

Gatun, one of the old *pueblos del rio* (river villages) was among the first communities to be relocated, because it was located at the site of the future Gatun Dam. In 1908, the community of six hundred people was disassembled one building at a time (an old church and schoolhouse, ten stores and a hundred homes) under the supervision of the canal administration and relocated to a site above the future lake. Thus, the *Gatuneros* were disconnected from the river—which had delivered travelers and money during transportation booms and silted floodplains for them to farm during bust periods—and reconnected to the urban infrastructure of the Canal Zone. The *Canal Record* described the community's relocation in this way:

The inhabitants of the village were notified some time ago to move their building but have paid no heed to the order until within the past few days when they were practically forced into action as the old site is now needed. They prefer their present location to the new one, it being on the river which they have used as their principal means of transportation. The new town will be amply provided with transportation routes, for besides being on the new line of the Panama railroad, it is connected with both the labor camp and the American village at Gatun by roads, and in the near future it will also be connected with Colon by road. It will be supplied with water and sewerage systems, advantages which the village did not possess.[38]

Rural land policy became more draconian as the waters of the Chagres River rose behind the dam. Canal authorities relocated dozens of villages, including construction work camps and *pueblos del rio* located upstream from Gatun, including: Ahorca Lagarto, Bohío Soldado, Frijol, Gorgona, and Matachin (figure 5.1). While many people in these communities had a long history in the area, others were canal laborers attempting to circumvent the control of the canal administration. Several years before, administrators had decreed that only laborers who purchased meal tickets—for food considered unpalatable and overpriced—could live in government housing. Five thousand workers subsequently moved to Panama City, Colón, or rural areas.[39] This population, dispersed along riverbanks and through forests, became a governmental problem for an administration that sought to control territory, workers' lives, and the spread of mosquito borne disease.

Figure 6.6
The railroad village of Bohio before it was flooded by Gatun Lake, January 1912.
Source: National Archives at College Park, Maryland, RG 185-G, Box 5, Vol. 9.

The Depopulation of the Canal Zone

A 1912 census listed the Canal Zone population as 61,279 and classified the enumerated as 30,948 blacks, 18,562 whites, 10,224 mixed, 648 Indians, 521 yellow, 373 Hindoos, and 2 Filipinos. By 1917, the population had fallen by half. According to the *Canal Record*, the thirty thousand people who emigrated from the Zone between 1912 and 1917 were largely "natives and colored people."[40] Why such a dramatic demographic shift?

A number of factors were at play as canal construction (1904–1914) came to a close. Many former laborers traveled home or migrated across Central America to work on other construction projects or in plantation agriculture. Those that remained on the isthmus were not allowed to settle freely along the Chagres River, as unemployed laborers did after the railroad and French canal projects in the nineteenth century. The US government ignored or encouraged rural settlement around transportation infrastructure during the first years of construction, but then things began to change.[41]

The depopulation of rural areas of the Canal Zone became policy when the American government made an eminent domain claim on all lands therein for canal purposes, even lands under private ownership. The US president, William Taft, decreed by executive order that "all land and land under water within the limits of the Canal Zone is necessary for the construction, maintenance, operation, sanitation, or protection of the Panama Canal, and to extinguish by agreement when advisable, all claims and titles to adverse claimants and occupants." If, according to the executive order, the US government and an occupant with a land claim or title were not able to reach an agreement, "the adverse claim or occupancy shall be disposed of and title thereto secured in the United States and compensation therefore fixed and paid in the manner provided in the aforesaid treaty with the Republic of Panama, or such modification of such treaty as may hereafter be made."[42]

The depopulation order of 1912 marked a major shift in policy because the entire Canal Zone, even private land, was suddenly declared "necessary" for canal purposes. It provided the legal basis for a project already underway. That same year, the Canal Zone established a rural police force to patrol a vast area crossed by rivers, tributary streams, and the rising lake. Eight officers were tasked with relocating "non-authorized" occupants from the Zone to Panama and preventing them from returning to their homes. They posted trespassing notices, visited houses and documented their residents, provided oral and written resettlement orders, and "assisted" inhabitants as they moved to sites above the level of the future lake.[43] The *Canal Record* described the response as follows: "Some moved promptly on being notified, others had to be taken out when the water was almost at their doors; some have disappeared in the higher silence of woods creatures; and some have moved in bodies, forming new settlements in which is preserved the community life of old."[44] Depopulation was more violent than the pastoral picture painted by this official communication organ. *The Workman*, a West Indian newspaper in Panama, characterized it as "ruthless destruction which was executed on the crops and shacks of squatters ... [with] poor and insignificant compensation."[45] The Zone police destroyed three hundred "native houses" to discourage resettlement in 1911 and early 1912.[46] By September 1913, an estimated 70 percent of the rural population had been removed from the Canal Zone.[47]

Rural enclosure distinguished the US canal project from the construction of the railroad and French canal. All three projects ended with the elimination of thousands of jobs, but the key distinction between the US canal project and its predecessors was the paucity of postconstruction options for the former work force. In the first two cases, unemployed laborers were able to make a living—however meager—by farming in rural areas along the river. However, when canal construction ended in 1914, the administration cut off access to nearby land. The enclosure of the Canal Zone—including agricultural landscapes along the Chagres and its tributaries—marked a transition from transportation as a sociotechnical project that could mesh with, and even benefit from, agricultural land use in the same region to a governmental project to produce a landscape that served narrowly defined purposes.

The US government compensated rural people able to provide official title to land. But, after the 1912 depopulation order, a difficult question faced canal administrators: Should so-called squatters, or people living in the Zone without title, also be compensated for the land they occupied? Some of these families had lived along the Chagres River for generations without formal land titles, but others—an estimated 50 percent at the time depopulation began—had migrated to the isthmus after 1903, mainly to work for the Panama Canal or the Panama Railroad.[48]

The Joint Land Commission, a body of US and Panamanian representatives that was created to adjudicate canal-related land and property issues, concluded that both titleholders and squatters had legal property rights under the 1882 Colombian Cultivators Law. This law was relevant to the case at hand because the lands in question were Colombian territory prior to Panamanian independence in 1903 and the subsequent creation of the Canal Zone. Because neither Panama nor the United States had passed laws that voided the established Cultivators Law, it remained in force under international law.[49] Thus, the US government was obligated to compensate "nonauthorized cultivators" (i.e., squatters) for their improvements (homes, crops, etc.), but farmers would be forcibly removed, nonetheless.[50]

The Canal Zone was an aquatic space: an artificial waterway provide the justification for its existence and the rivers and reservoirs that fed that waterway drained much of its area and shaped its territorial jurisdiction. The Canal Zone extended out five miles from the shipping lane on either

Figure 6.7
Canal Zone police force, 1913. *Source:* National Archives at College Park, Maryland, RG 185-G, Box 9, Vol. 17.

side (again, it was ten miles wide in all) and further into Panama where the waters of the lake exceeded the limit. In such cases, the Zone-Panama boundary was set at the hundred-foot contour—fifteen feet above the lake's eighty-five-foot operating level (figure 6.2). Yet, before the lake was full, the topographical contours that demarcated the boundary between the Canal Zone and Panama were practically unmarked.

The new limits of US jurisdiction (the hundred-foot contour above sea level) were only visible on the maps that Zone police carried into the field and through a few scattered boundary markers. There were no border fences, so state power operated in more imprecise ways on the ground than in official decrees. Police reportedly showed relocated people "about" where the boundary was located, as they explained the flooding of Gatun Lake and depopulation order. In some cases, residents questioned the imminent rise of the lake. The *Canal Record* reported, "It is difficult to persuade some of the inhabitants that the inundation will ever take place. One old bush settler, after receiving repeated warnings heedlessly, ventured it as his opinion that the Lord had promised never again to flood the earth."[51] The depopulation gave rise to small acts of resistance and outright dissent. For example, police reported "malicious interference" with new Zone boundary markers, including removal and defacement.[52] They also struggled to remove those who resettled on the new lake's islands and banks, refusing to move beyond the hundred-foot contour.

Limón and the Limits of Rural Government

On October 10, 1913, the recently elected president, Woodrow Wilson, pressed a button in Washington, D.C. that triggered a dynamite blast and destroyed the Gamboa Dike. Water from Gatun Lake flooded into the Culebra Cut and the canal was complete. The US government had rerouted the Atlantic-bound Chagres to the Pacific through the continental divide. The canal builders—a cast of thousands including engineers, administrators, scientists, and workers—had achieved the old dream of building a waterway across the isthmus ahead of schedule and under budget.

Yet even as people worldwide celebrated a great engineering achievement, the territorial problems around the waterway had just begun. As the story of Limón—a river town near the confluence of the Chagres and Gatun Rivers (figure 5.1) illustrated, the reorganization of land, water, and society across hundreds of square miles was a difficult, drawn-out, and politically charged process. A letter written by Canal Zone administrators narrated the community's story in this way:

The village of old Limón was located on the Gatun River on about the 55-contour line in the vicinity of the present bascule bridge at Monte Lirio. The inhabitants there were required to move, after being fully compensated by the Isthmian Canal Commission for their improvements. The people at old Limón, without consulting the Canal authorities, selected another site for their town. The new site was in the lake area. When the settlers commenced to build houses there they were advised by the agents of the Canal that the location chosen was within the lake area and that they could not remain there. Those who had already made improvements were compensated for them, although there was no obligation on the part of the United States Government to pay them anything. The Isthmian Canal Commission then suggested a new site for the old town of Limón on the north side of the river, but the Panamans went to another place, situated on an island in the lake, with which selection the Canal authorities had nothing to do. This necessitated another removal, and Governor Goethals assisted the people very materially in moving to their present location. The new site was cleared for them and considerable grading was done in order to put the ground in suitable condition for building. The Canal took down most of the houses, transferred the material, and rebuilt the houses at its own expense and without cost to the settlers.[53]

When Gatun Lake reached its operating level in 1914, it was the largest artificial reservoir in the world, covering 164 square miles of the Chagres River valley. That same year, government representatives from Colón and the Panama Canal Company selected a location for Limón on a peninsula in the Republic of Panama on the north bank of the Gatun River just above the hundred-foot contour. The account of Limón presented by canal administrators adopted a tone of patronizing beneficence, but, for some Panamanian observers, the story underlined the human cost of reengineering the isthmus. In a 1914 newspaper article entitled "A voice crying in the wilderness?," a Panamanian journalist used biblical allusions to recast the people of Limón not as squatters, but a nomadic tribe dispossessed of its land—"poor countrymen ... denied the benefit of the land of their birth, which they have watered with the abundant sweat of the brow and the fruits of which they require to support their offspring."[54]

The Society of the Chagres, a social club for elite white Canal Zone men, also invoked the desert as they joked, rather callously, about the intractable land conflicts around the waterway in a "hymn" written for a 1916 meeting to the tune of a popular song of the day:

HYMN 1932 A.D.

TILL THE LANDS OF THE SQUATTERS GROW COLD

(Sung by Judge Feuille and the Joint Land Commission. Tune "Till the Sands of the Desert Grow Cold.")

Till the lands of the squatters grow cold,

And the infinite claimants are old,

We'll scrap endlessly,

No truce shall there be,

Though lawyers may threaten and scold,

Till the Paymaster runs out of gold,

And the mysteries of law shall unfold,

We'll cling, job, to thee

And draw salar-e-e

Till the lands of the squatters grow cold.[55]

In the end, the canal company created two large reservoirs, drowning fifty communities and flooding two hundred square miles of land.[56] As the Chagres River was reorganized to move ships through a series of downstream locks, water management infrastructure was extended further upstream. Gatun Lake was the largest artificial lake in the world in 1914, but even then engineers recognized that its storage capacity would be insufficient as traffic increased.[57] Thus, US President Calvin Coolidge signed an executive order in February 1922 to create a second dam and reservoir on the upper Chagres River. The 22 square miles that would become Madden Lake were expropriated from Panama, again under the terms of the 1903 canal treaty, and appended to the Canal Zone. The territorial politics enacted on the lower Chagres River at Gatun—survey, expropriate, depopulate, and flood—were repeated on the upper Chagres, initiating a search for more fresh water that continues today.

The Clearing: From Territory to Biopolitics

The Chagres River and its tributaries were reorganized according to the water demands imposed by the lock design, the climatic and hydrological specificities of Panama, and changing volume of traffic passing through the canal, but water management was also a technopolitical project that linked civil engineering and governance. The transformation of physical geography to supply canal water precipitated the reorganization of the Chagres River basin's human geography according to racial and national categories, beginning—but not ending—with the forced resettlement of West Indian and "native" residents from rural areas in the Zone.

In the most obvious sense, the untold human story of creating the Panama Canal waterscape was dispossession. Thousands of rural people in Panama were forcefully removed from the land in the name of progress, modernization, and development, like others living along dammed rivers worldwide in the twentieth century. But the governmental rationales for rural depopulation in the Canal Zone were more multifaceted than this straightforward narrative suggests. Water was not simply an unmediated resource and object of political control; it was imbricated in the socio-technical networks through which power was deployed around the canal.

For administrators, depopulation was a means of *clearing* the Canal Zone to establish space for still undefined transportation infrastructure and render the heterogeneous groups of people scattered across the floodplain of the Chagres River more legible and governable by relocating them to settlements and cities. Thus, water management had both territorial and biopolitical dimensions. Canal administrators managed water for shipping and to control threats to the security and health of white US populations.

Much of the length of the Canal Zone-Panama border was unmarked and, even where boundary markers existed, flowing water and flying mosquitoes did not recognize their authority. The Department of Sanitation advocated for depopulation to reduce the risk of disease in Canal Zone communities by moving rural people into Panama—either to cities or planned resettlement communities in the countryside. Administrators saw autonomous communities as sanitary problems because, among other reasons, they were often full of standing water—barrels, latrines, and puddles—that provided breeding grounds for mosquitoes that might carry malaria and yellow fever to nearby white communities. However, the ecological irony of sanitary policies was that canal construction itself produced "unsanitary" landscapes conducive to malaria and yellow fever outbreaks by bringing together tens of thousands of foreign human bodies without immunities, while excavation work and reservoir flooding created bodies of standing water full of vegetation where mosquitoes thrived. "Every time a steam shovel made a deep hole," observed the wife of canal sanitation chief William Gorgas, "water would almost immediately collect, and the Anopheles [mosquitoes] would at once seek such a depression for a breeding ground."[58]

Sanitary water, by contrast, was understood to be liquid that was covered and circulated by urban infrastructure. Thus, the US government built modern water and sewage systems in both Panama City and Colón in return for relatively small annual payments from the government of Panama. In other words, these infrastructures—which still circulate water from canal reservoirs to the terminal cities today—were not simply built for development, but as a component of a regional sanitation campaign.[59]

Figure 6.8
Water infrastructure built for drainage and mosquito control in the community of Paraiso, Canal Zone. *Source:* National Archives at College Park, Maryland, RG 185-G, Box 10, Vol. 20.

Water and sewer infrastructures established a transboundary definition of citizenship. The Zone government integrated forced rural depopulation and extensive urban infrastructure to make human ecologies more legible. As in other colonial and imperial settings, technical networks and governmental techniques of surveillance extended state control of subjects without political representation.[60] Rural settlement and agricultural production were seen as undesirable from a labor management perspective, because administrators—invoking colonial discourses of tropical fecundity and racialized laziness—worried that "negro" laborers might desert wage labor for the supposedly "easy life in the bush."[61] Beginning with the relocation of the village of Gatun from the banks of the river in 1908, Panamanians and former West Indian laborers entered into new relationships with the Canal Zone and US government mediated by state functionaries and also—crucially—the pipes, sewers, reservoirs, and paved roads that radiated out from the Zone enclave into Panama.

Conclusion: From Engineering to Governance

The phrase "let the dirt fly," often associated with the construction of the Panama Canal, recalls a modernist faith in cultural and technological progress that is troubled in the twenty-first century. If earth was the element that represented the attitude of conquest that defined modern humans' relationship to nature at the turn of the last century, then water is the element that reveals how we live with the legacies of that era. The excavation of the canal could be finished, but dredging sediment from its channel and providing a constant water supply to the so-called big ditch is never complete. Such maintenance work is formatted by technical decisions made a century ago and the increasing demands of global transportation.

As the construction of the Panama Canal approached completion in 1914, the most pressing questions for administrators shifted from the sphere of engineering to governance. Which—and how much—land was necessary for canal purposes? Who—if anybody—would be allowed to occupy and use lands not immediately necessary for those purposes? These questions were difficult to answer when the future of shipping and role of the Canal Zone in US imperial ambitions were both unknown. The public rationale for rural depopulation was to establish a dedicated space for

transportation and to improve sanitary control, but the scope of new policies allowed administrators to clear human settlement—largely black West
Indian former laborers and their families—in anticipation of undefined
future projects. Thus, the depopulation of the Canal Zone was part of a
broader bioterritorial effort to buffer white populations and infrastructure
from human and environmental threats that were, at least in part, byproducts of the construction of the canal itself.

But this is not only a story of control, because social life—like water
itself—confounds the best-laid plans. Canal engineers designed technologies intended to discipline the Chagres River for navigation and policies
to relocate people across its floodplain, but locks and dams also interacted
with unknown or unacknowledged social and ecological relationships,
often producing unintended consequences. Among these was the emergence of Gatun Lake as a "highway" for both oceangoing ships and the
canoes and launches of rural people from relocated communities like
Limón who were pursuing new opportunities opened up through the
emergence of a new and expansive water world.

Figure 7.1
Cayuco (canoe) on the banks of Gatun Lake in Limón. Photo by the author.

I first visited Limón in March 2008. That morning, I caught a bus at a busy shopping mall on the outskirts of Panama City. Not realizing that I had mistakenly boarded a "regular" bus that would crawl through miles of urban sprawl before reaching the highway (rather than a more direct "express" bus), I wedged my knees into the torn brown vinyl seatback in front of me. A long three hours later, I stepped down onto the shoulder of the Transístmica, the congested highway between Panama City and Colón that runs parallel to the canal, and caught a taxi into town. It was late in the dry season and the air was hot, dusty, and yellow.

Limón—a town with approximately six hundred residents, nearly all of African descent—is located on a peninsula where the Gatun River flows into Gatun Lake.[1] The peninsula is covered by a couple hundred wooden

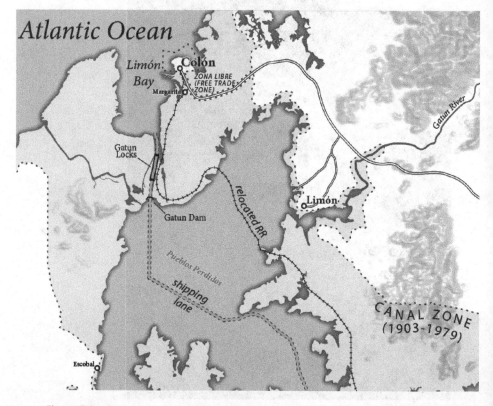

Figure 7.2
Location of Limón on Gatun Lake, outside the former Canal Zone. Image by Tim Stallman, used with permission.

and cinder block homes that are painted in peaches, pinks, and blues and crowned by the steeple of a Catholic church. Limón's history is bound up with that of the Panama Canal. It was founded in 1914, the year that the lake reached its full operating level and the first ship passed through the locks. Or, more accurately, the canal administration relocated the people from the riverside community of Limón Viejo (old Limón)—now beneath the waters of Gatun Lake—to the peninsula that year.

I chose the community as an ethnographic field site because I wanted to better understand the changing relationships between the canal administration and the people living in the surrounding region. Before the transfer of the Canal Zone from the United States to Panama, which took place between 1979 and 1999, the lives of people in Limón were connected to the enclave, where many of the men worked on landscape maintenance crews cutting grass and cleaning roadsides and women worked as maids or, in some cases, did laundry and other piecemeal work for soldiers and canal employees. However, the community's historical participation in the transport service economy runs deeper. Residents trace their family histories to laborers who migrated to the area to work on the Panama Railroad (1850–1855), French canal project (1881–1889), or US canal project (1904–1914). And yet, the older *Limóneses* told me in interviews that the best times were not construction booms, but periods when agricultural production thrived—particularly the era when the banana business infused cash into the lake region between the 1920s and the dawn of the Second World War (chapter 8). The war precipitated a new round of infrastructure construction projects around the canal (new locks, extended electrical and road networks, military quarters, radar stations, antiaircraft batteries, and camps), which required large labor forces recruited from across Panama, Central America, and the Caribbean. Because the labor demands were large and projects urgent, the wages were relatively high.

Employment opportunities have become scarce for the poor and poorly educated in a region that is politically, economically, and geographically organized around a transport service sector that, both globally and domestically, requires fewer laborers than it used to.[2] For most, agrarian livelihoods are untenable or undesirable and there is little regular work in the community. Those with jobs—which is to say, those who are not retired and drawing pensions, unemployed, or piecing together day labor—commute to Colón. Most of the commuters have found employment as low-paid manual laborers at the city's ports, transshipment terminal, or free trade zone.

Before the completion of the Transístmica highway in 1943, Limón was part of a water world (chapter 10). The river and the lake were the main thoroughfares. As large cargo ships transited the canal via the shipping lane in the deepest part of the lake, launches (large open motorboats) circulated around its banks, carrying passengers between a dozen lakeside communities and Colón. For people in Limón, the launches—La Bruja, La Faustina, Leilei, El Sol No. 1, El Sol No. 2, La Rosita—were a lifeline to money, goods, and services. In the absence of paved roads that could be traveled throughout the long rainy season, boats were the only way to get agricultural products to market, shop for necessities, and transport the community's sick and injured to receive medical care. After the war, it became common for a new generation—especially men—to take a wage job in the Zone, while growing rice, beans, corn, and tuber crops in *montes clandestinos* (secret farm plots) hidden in the forest of the Zone.

The flooded river has a diaspora: a ring of towns whose founders, like those of Limón, were resettled above the hundred-foot contour during the creation of the lake. I met Paulo Ortega Escobal, a town perched on the west bank of the lake directly opposite Limón (figure 7.2). When I walked up to his house, Paulo was sitting on the front porch. "Canal Protection Division," he said, joking, by way of introduction. He offered me his hand and then presented three laminated canal identification cards—one at a time—with photos of a younger, more serious-faced version of himself. A wooden plaque on the wall honored his thirty-two years of service to the canal, mostly as a security guard at the Gatun Locks. Paulo's quick identification with the Canal Zone made me reflect on how I—the tall *gringo* wandering inexplicably into town—must have looked to him and others in a place like Escobal that had such deeply ambivalent relationships with people who looked and spoke like me in times now past.

Paulo grew up in Escobal, where US soldiers from nearby military bases were regular visitors. At night, they filled the local cantinas to drink and dance, including a popular one called *El Dump*. Paulo moved away as a young man and spent his working years in Colón. He, like his peers in Limón, was part of the postwar generation that left the lake and farming behind for city work. He moved home when he retired and began to receive a pension from the US government, which makes him somewhat wealthy by the standards of the lakeside community.

Paulo told me that Escobal was among the towns relocated during the creation of the lake. A number of the community's founders came from Limón Viejo (old Limón) (figure 5.1). Some of that town's inhabitants moved to the lake's west bank, where they founded Escobal. Others (were) moved to the east bank, where they founded a new Limón. So the towns are "brothers," he said.

After an hour, I thanked Paulo for his time, shook his hand, and wandered through the town square, where billboards announced government infrastructure projects: five hundred and fourteen thousand dollars to fix the road and fifty-two thousand dollars to repair a water pipe. Fifteen minutes later, I was eating in a small restaurant off the square when Paulo's granddaughter ran up to my table and handed me a scrap of paper with a scrawled note from him:

CUANDO EL
LAGO GATUN
SE FUNDO QUEDARON
EN LA PROFUNDIDAD
DEL LAGO
SON MONTE VISTOSO
CALLE BRUJA
LIMÓN VIEJO
SAN JUAN
Y MONTE LIRIO

"When the lake was created, Monte Vistoso, Calle Bruja, old Limón, San Juan, and Monte Lirio remained in its depths." Back in Panama City, I filed the note away, but I was struck by the fact that Paulo considered those historical villages—their names, their common fate, and their postdiluvian connections—important enough to send his granddaughter running after me with the note. Before my trip to Escobal, I had not thought much about the world lost beneath Gatun Lake.

Gil Blas Tejeira popularized the term *pueblos perdidos* (lost towns) in Panama in his historical novel by that name. In the book, Tejeira—who worked as a journalist, teacher, librarian, provincial treasurer, and ambassador during his lifetime—captures the rhythms of everyday life and language of the

nationally, racially, and linguistically diverse groups that intersected during
the canal construction era. He provides a textured, intimate account of
that time and reminds us that those displaced by the creation of the waterway were not a homogeneous population of squatters or workers, but real
people who inhabited rich social worlds and had their own histories,
dramas, and aspirations.

Nobody alive in Limón today ever laid eyes on the *pueblos perdidos*. But,
in oral history interviews, old men and women recounted stories of flooding, migration, and settlement passed down from previous generations.
While telling these stories, some pointed out the spot on the lake's surface
above the townsite of old Limón and to the hilltop-turned-island where
villagers—pursued by Zone police—stayed for a time as the water rose
behind Gatun Dam before settling on the peninsula in 1914. They talked
about river life and the faraway places—Cartagena, Colombia, Caribbean
islands, France, even Africa—that their ancestors left behind for Panama
and construction work on the railroad or canal.

Old Limón was a few miles upstream from Gatun, one of the most
important Chagres River communities because it was a stopping point for
boat travelers, the location of a Panama Railroad stop, and a construction
center for the French canal project. Tejeira described the town before the
flooding as follows:

The houses of Gatun were small and covered with zinc roofs, shaded by fruit trees
and coconut palms. A church of wood and zinc was erected on a cement platform
with steps. Perhaps because the bell tower, also of wood, was not strong enough to
support the weight of the bells, the two which, according to the *pueblo*, had been
hung by a beam supported by two strong pillars. The economy of the *gatuneros*
depended primarily on the bananas cultivated on the banks of the Chagres and its
immediate tributaries ... There were few whites, many mulattos, and mostly blacks,
all well integrated.[3]

The idea of *pueblos perdidos* resonates for reasons beyond the trauma of
historical events themselves. The flooding of the canal and depopulation
of the Canal Zone landscape became the foundation for a cluster of stories,
emotions, and symbols—a way in which people have assigned meaning to
ambivalent attachments to transportation projects.[4] The flooding took
place a century ago, so memories are more like Impressionist watercolors
than realist landscape paintings. One woman remembered being told as a
child that, as water climbed the trunks of trees still covered in leaves,

people from the river valley paddled their canoes beneath the green canopy in awe of the transformation of the landscape.

Even with the flood, all was not lost, because the lake connected communities with a shared history of displacement. Flotsam from the riparian world—plants, animals, building materials, and cultural traditions—washed up on the lakeshore. For example, Tejeira wrote that every May, old Gatun filled up with pilgrims who came to celebrate the town's patron saint, Santa Rita. They came from towns along the river and train line, Colón, and the *costa arriba* and *costa abajo* (upper and lower Caribbean coast) for a church procession, as well as cockfights, bowling, and dancing that could last for an entire week.[5]

Today, Limón's patron saint is also Santa Rita. Like in old Gatun, she is honored every May with a procession, dancing, and music. The official festivities (but not the long nights of drinking and thumping Panamanian reggae) are organized by the Catholic church, Santa Rita de Castilla, which was built with community contributions during the first decade after the

Figure 7.3
Family photos from the canal construction era in Limón. Photo by the author.

flooding. One woman told me that when she was a child, "The people came from far away and stayed for days. There were cars parked from my house down to the end of the street on both sides for the fiesta and the dances. The dances got so full!"[6] The Santa Rita festival illustrates a cultural connection to geography and historical experience that is also manifest in a language, housing, and food.[7] A century later, Limón residents retain connections to histories and landscapes submerged beneath the lake. Thus, the *pueblos perdidos* are physically "lost" and culturally present.

The phrase "lost villages" is also an apt characterization of how an increasingly mechanized, intermodal global transportation infrastructure less dependent on manual labor has left places like Limón behind. From the Second World War through the 1970s, many men and women from Gatun Lake towns worked as laborers in the Canal Zone. A number of older men told me that they worked for the Maintenance Department as *macheteros* (machete men) cutting fast-growing vegetation in Canal Zone communities and military forts and along roadsides. Many of their wives, mothers, and sisters did laundry, cooking, and domestic work in the Zone. During the transfer of the canal from the United States to Panama, most of them lost their jobs. The layoffs were a side effect of the change in canal administration as the Zone was hollowed out and its large population of US citizens—who hired large numbers of Panamanians to do domestic work—dispersed. But people in Limón said that cronyism played a role, too, claiming that the new Panamanian administrators selectively hired their family members and friends. Thus, for community members of a certain age, the era of the *gringos* is remembered with bittersweet, postcolonial nostalgia, as a time of greater access to jobs and social services marred by insults to national pride and everyday (often racist) individual indignities.

In Limón, people analyzed the community's changing relationships with the Panama Canal through the metaphor of landscape maintenance, specifically *limpia* (clean) and "*sucia*" (dirty) landscapes. Some said that the current administration, the Panama Canal Authority, has allowed the area around the waterway to become "*sucia*," which they contrasted with the "*limpia*" landscapes of the former Canal Zone. The use of these adjectives to describe landscapes is common in Panamanian agriculture, where clearing and weeding land by machete is called *limpiando* (cleaning) and areas considered overgrown are often labeled *sucia*. Thus, when describing an open or cleared landscape, rural Panamanians often refer to it as *"limpia"* and, approvingly, as *"bonita"* (attractive).

In my interviews in Limón, it was common for older residents to describe historical Canal Zone landscapes as both *"limpia"* and *"bonita."* On the surface, this is not surprising given the landscape aesthetic of short-cut grass associated with the canal's antimosquito sanitation campaigns. However, the clean/dirty distinction also provides people with a means of talking about the present in terms of the past. "The lake was clean," Eneida, a lifelong resident of Limón, told me, "not like today with the ugly [weedy] banks, all of the lake had clean [mowed] banks." She, like many others I interviewed, was quick to point out that the Canal Zone, the Gatun Locks, and banks of the lake had become overgrown since the *gringos* left. She continued, "The people respected the laws of the Zone. If you threw trash on the side of the road, there were fines. But the Panamanian government," Eneida said, laughing, "doesn't *know* the word maintenance."

Framed in this way, weediness signals the abdication of responsibility to place and community. "How can you live somewhere and not keep it clean?" Eneida asked me as she finished making her point. Without maintenance, she pointed out, weeds choke the productive landscapes that people have worked hard to establish—places invested with identity and meaning. To talk about the canal in such language flags the perception that the transport service economy has abandoned not only yards, roadsides, and buildings, but people as well. While the opening of the canal entailed social disconnection through the flooding of the river and forced resettlement, weediness reflects disconnection through neglect and the disregard of maintenance and responsibility.

Disconnection is different from never having been connected at all. Anthropologist James Ferguson writes, "Disconnection, like connection, implies a relation and not the absence of a relation. Dependency theorists once usefully distinguished between a state of being undeveloped (an original condition) and a state of being underdeveloped (the historical result of an active process of underdevelopment). In a parallel fashion, we might usefully distinguish between being unconnected (an original condition) and being disconnected (the historical result of an active process of disconnection) ... [what this reveals] about globalization is just how important disconnection is to a 'new world order' that insistently presents itself as a phenomenon of pure connection."[8] As people in Limón point

out, weeds thrive in spaces of global disconnection where capital cleared
and then maintained a "clean" landscape before moving on.

If the *pueblos perdidos* were "lost" during the flooding of Gatun Lake,
then Limón and other communities have been orphaned by the canal and
the transport economy over the past several decades.[9] Limón was never a
self-sufficient "local" rural community impacted by global forces. It grew
up at the watery margins of an emerging global infrastructure that its
inhabitants worked to build and maintain. Despite the everyday affronts
to personal and national pride associated with working in an imperial
enclave, many community members in Limón had jobs as full-time laborers within the state-subsidized, residential US canal project. The community's place in Panama's contemporary neoliberal economic paradigm is
arguably more tenuous. These two types of *pueblos perdidos*—one historical,
one contemporary—illustrate the argument that infrastructures rework
landscapes in a manner that facilitates new connections and opportunities,
but, also creates barriers and forecloses alternative possibilities.

8 The Agricultural Possibilities of the Canal Zone

Figure 8.1
Ship passes through the canal with banana plants in foreground. *Source:* National Archives at College Park, Maryland, RG 185-G, Box 6, Vol. 11.

Here in the most important country of its size in the world, the Panama Canal is meeting an unprecedented problem: in three hundred twenty-five square miles of moderately suitable territory, more or less isolated, how nearly self-supporting shall this area be made, what sort of horticultural development should it have and how much?

—Otis W. Barrett, Canal Zone horticulturalist, 1915[1]

In 1909, five years after canal construction began, the soil scientist Hugh H. Bennett and horticulturalist William A. Taylor arrived in Panama to study the agricultural possibilities of the Canal Zone. At the request of George Goethals, chief engineer of the canal the US Department of Agriculture sent the scientists to the isthmus at the height of construction to explore the potential of the country's new tropical possession for gardens, farms, and ranches. Goethals's request was, at least in part, a response to growing complaints and concerns about the canned and shelf-stable provisions that administrators fed tens of thousands of workers and concerns about the lack of fresh produce.[2] But, more importantly, Bennett and Taylor's agricultural study engaged a number of emerging questions about what kind of space the Zone was and could be.

The Canal Zone has often been represented as a massive construction site that was quickly transformed into a planned and manicured US suburb in the tropics after the waterway opened in 1914.[3] In fact, the enclave was largely rural, practically roadless, and poorly surveyed in the decades following the first transit. State authority was limited and unstable across this weedy, postconstruction landscape. Actors within and beyond the canal administration debated plans for the future of hundreds of square miles of Canal Zone land that were not immediately necessary for transportation or residential purposes. Yet we know remarkably little about the history of these rural lands.

The Zone's rural question—how to manage the landscapes depopulated by the US government beginning in 1912—had implications beyond land use per se, because it raised contentious questions about the scope of US territorial ambitions in Panama. Were canal administrators simply operating a navigable waterway or were they constructing a quasi-autonomous imperial enclave? The parameters of the canal project were thus under negotiation in Washington, D.C., Panama City, and Balboa, site of the headquarters of the canal company. But they were also being worked out on the ground across the Zone's fields, forests, rivers, and lakes.

The question of *how* to use rural lands implied a critical second question: *who* (if anyone) should use them? In 1914, the year the canal opened, the commission laid off an astounding thirty-eight thousand employees, mostly black West Indians. Many returned home or migrated elsewhere, but forty to fifty thousand West Indians (workers, their families, and many others) remained in Panama, igniting racial, cultural, and political tensions by driving up rents and competing for jobs. Within the combustible context of urban unemployment and unoccupied rural land, the US government reversed the depopulation policy and resettled the Zone.

The policy shift led to an unexpected outcome: a banana boom around the canal. In 1922, the Canal Zone government initiated a program through which former laborers could apply to lease small farm plots in the Zone. This relief program and social experiment transformed the landscape around the canal in ways that made some white administrators uncomfortable. The lessees were predominantly black, lived in autonomous settlements, and planted bananas everywhere. These settlements were, in other words, different from white, strictly regimented, and highly planned Canal Zone communities in almost every way. The banana boom was an unexpected outgrowth of canal infrastructure within the demographic, cultural, political, and ecological context of the isthmus.

During the *auge* (boom) of the 1920s, farmers, *compradores* (middlemen), state functionaries, and capitalists came together to negotiate the production, transport, and sale of "green gold" around the canal. When the boom peaked in 1927, 2.8 million bunches of bananas were shipped from Canal Zone ports. While this did not approach the exports from Latin America's so-called banana republics, led by Honduras, it was large given the Zone's limited area, dominance of smallholder production, and unique political configuration.[4] Thus, the banana's history in the Canal Zone grounds a transportation project firmly in place, revealing complex connections among politics, ecology, and infrastructure. By drawing our attention to the racialized history of land use planning and struggles around the canal, this iconic fruit reveals the possibilities and problems of farming on the banks of a waterway that was built for navigation purposes.

The Banana as an Infrastructural Species

Modern bananas might be described as an *infrastructural species*. The commercial varieties raised for human consumption are sterile and do not produce seeds. The plants, which grow from underground stems called rhizomes (or corms), are propagated by removing buds (or suckers) from the corms of existing plants and replanting them. Modern bananas are an infrastructural species because they largely depend on human labor for propagation and transportation infrastructure for dispersal. As a result, they thrive along edge environments where transportation networks and lowland tropical ecology meet. Thus, the nature-culture borderlands of the Canal Zone—with its waterways, alluvial soils, train line, heavy rainfall, and nearby ports—provided ideal habitat for bananas.

Figure 8.2
Bananas along the Gaillard Highway, Canal Zone, 1922. *Source:* National Archives at College Park, Maryland, RG 185-G, Box 7, Vol. 14.

Meanwhile, in Central America's so-called banana republics, including the Republic of Panama, large-scale monoculture production was held up as a path to economic development, so governments actively created the political, ecological, and infrastructural conditions for banana production and export to thrive. The Canal Zone, by contrast, was an imperial enclave organized around transportation. Although its government had access to a large workforce (former construction laborers) and unused land (areas depopulated beginning in 1912), administrators disagreed about the costs and benefits of promoting agriculture. For some white canal administrators, banana cultivation had positive connotations (self-sufficiency and poverty relief), but, for others, the fruit had associations (backwardness and blackness) deemed undesirable in the Zone.

Bananas were—and remain—plants of empire. Botanists believe that the wild ancestor of the modern banana was domesticated in highland New Guinea around 5,000 B.C. Edible varieties spread across the Old World tropics and accompanied European colonists as they arrived in the Caribbean.[5] In *Natural History of the West Indies*, Gonzalo Fernández de Oviedo reported that colonists introduced bananas to the neotropics in 1516. He wrote that, after a decade, the plants had "multiplied so greatly that it is marvelous to see the great abundance of them on the islands and in Tierra Firme where the Christians have settled."[6]

The overlapping geographies of colonialism and bananas are not surprising. As the historian Alfred Crosby has shown, colonial conquest was a collective effort by networks of Old World biota (dominated by European humans) that evolved in conflict and cooperation.[7] The banana plant, a tropical perennial, was not technically part of European invasive networks of humans, grasses, and livestock, but its rhizomes—transferred for propagation—traveled with them, benefitting from and supporting the ecological reorganization of the neotropics.

If bananas were plants of empire, they were also plants of revolution and subsistence for runaway slaves and laborers who built transportation networks and worked on the plantations that undergirded extractive political economies throughout Central America and the Caribbean. To follow the historical circulation of banana rhizomes, then, is to retrace transportation routes that open up complex cultural encounters and political ecologies of colonialism and imperialism.[8]

During the nineteenth and early-twentieth centuries, railroad lines were built across Central America and large swaths of adjacent forest cleared and planted with banana monocultures. The origin myth of banana capitalism illustrates how the expansion of transportation networks and banana production were mutually reinforcing—even coextensive—territorial projects. The story goes like this: Minor Keith, a businessman from New York by way of Texas, began his career in the shipping business

like many of the banana capitalists of his day.[9] He traveled to Costa Rica
in 1871 to work for his uncle, Henry Meiggs, then under contract with its
government to build a railroad between San Jose and the Caribbean port
city of Limón. Keith took charge after Meiggs passed away in 1877. After
difficulties with Chinese and Italian workers who were unwilling to endure
yellow fever and malaria for low wages, he recruited heavily in the West
Indies, particularly Jamaica, where—incidentally—commercial banana
production dated to the early nineteenth century.[10]

When the Costa Rican government defaulted on payments to Keith
and construction stopped, he offered the idle West Indian work force
small plots in the riparian bottomlands along the tracks to grow food
so they would not abandon the project. According to legend, he soon
discovered that the workers were cultivating bananas in these "provision
grounds" and realized that the fruit might become profitable export freight
for the unfinished railroad. Costa Rica later gave Keith eight hundred
thousand acres along the tracks as payment, and he started planting
bananas.[11]

From that point on, banana production and railroad construction
were inextricable in Central America, but the process of expansion varied
across countries. In Costa Rica and Guatemala, for example, railroad
lines were built first between major cities and ports, and bananas followed the tracks after they were laid. Elsewhere, particularly in Honduras, railroads traversed the fruit districts first and left the cities
unconnected. Therefore, the landscapes that emerged were neither tech-
nologically preformatted (railroads producing homogeneous tropical
space), nor geographically predetermined (responding to the "natural
advantages" of the tropics), but the hybrid and contingent products of
a technoecological network extending across political, economic, and
cultural difference. If Guatemala was not Honduras, then the Canal Zone
was neither.[12]

As we've seen, the geographical advantages of the Panamanian isthmus
as a transit zone were reinforced through the historical layering of built
infrastructure (roads, a railroad, and canal projects) requiring large labor
forces, which Panama historically lacked (chapter 5). Thus, path-builders
recruited laborers elsewhere—like Minor Keith, many found poor Caribbean locales to be the best recruiting grounds—a practice which precipitated large in- and out-migrations that reshaped regional demography.

Along the Chagres River—like Keith's railroad at a larger scale—the uneven availability of construction jobs over time meant that workers moved between wage labor and farming as socioeconomic conditions changed. Agriculture provided rural people with a means of mitigating the livelihood risks associated with the vagaries of the transport service economy.

Tensions around Food and Land Use in the Canal Zone

The US Department of Agriculture scientists Bennett and Taylor arrived in Panama in November 1909 to study the possibilities for agriculture and food production within the Canal Zone. They described the food supply as follows: "Canned fruits, vegetables, fish, meats, and butter are largely used, though these more perishable products are supplied in large quantities in the fresh state from the United States. This is accomplished by shipment in refrigeration from New York and New Orleans to Colon, at which point such articles as require it are placed in cold storage, where they are available for daily distribution by supply trains across the Isthmus as required."[13]

Feeding an enormous labor force had been a persistent problem for administrators intent on completing construction. The logistics of moving calories in bulk from agricultural areas to construction sites was only the supply side of the food problem. How workers ate was also shaped by their cultural backgrounds and preferences. Despite the availability of affordable meals in mess halls, some West Indian laborers were reluctant to spend wages on food and skimped on eating to save money. This alarmed administrators concerned about the productivity of the largest group in the labor force by ethnicity. Laborers' relationships with food were irreducible to money. Many West Indians were accustomed to growing their own food, which administrators discouraged.[14] Meanwhile, for white North American employees, food gripes often turned on the quality and nutrition of the offerings. "Too much canned food and solid diet is being used at the present time," the chief sanitary inspector wrote Colonel William Gorgas, head of the canal's Department of Sanitation, in 1906. He went on to say, "Green vegetables are very scarce and high priced at most places in the Canal Zone, and at many places cannot be had. ... I am also informed by various District Physicians that it would be very advantageous for the physical condition of the ICC [canal commission] employees and laborers to have sufficient vegetables available at a reasonable price so that the employees would not be compelled to use canned stuffs daily."[15]

Rural land use, like food, became more politicized as construction proceeded. In a letter to the secretary of war in 1909, Goethals expressed enthusiasm about growing food for workers on rural land away from Canal Zone communities and canal installations. "These lands," he wrote, "derive practically their entire value from their agricultural possibilities."[16] But he and other administrators also questioned whether agriculture was a suitable use of canal appropriations. In a letter to a subordinate the same day, Goethals expressed ambivalence about the Zone's agricultural possibilities and concluded that farming was too tangential to construction to be funded, but should be handled delicately. He wrote, "Sentimental interest attaches itself to the development of the agricultural possibilities of the Zone that must be taken into consideration, and if we dropped this work entirely it would put us continually on the defensive and possibly subject us to some unfavorable criticism."[17] Goethals recognized, presciently, the moral-economic force at the nexus of food, land, and labor.[18]

Figure 8.3
"Silver roll" school garden, Empire, Canal Zone, 1910. *Source:* National Archives at College Park, Maryland, RG 185-G, Box 9, Vol. 17.

Early efforts to develop Canal Zone agriculture began, haltingly, through horticultural experimentation and land lease programs. The first horticulturalist employed in the Canal Zone, Henry Schultz, was hired by the Department of Sanitation upon the recommendation of well-known US Department of Agriculture "plant explorer" David Fairchild in 1906.[19] Fairchild gushed to William Gorgas, Schultz's new boss, "You doubtless realize more fully than I do the tremendous landscape possibilities which the Canal Zone affords. No other great shipping canal lies in a region where tropical palms, bamboos, and the thousand and one gorgeous tropical trees can be grown, and if properly started there is no reason why, by the time the canal is finished, its banks should not be lined with specimens of the most gorgeous tropical palms, bamboos and foliage plants in the world."[20]

The Department of Sanitation was not interested in food production. It sought to develop sanitary and "homelike" landscapes appealing to US Canal Zone residents. Henry Schultz, who began as a landscape gardener on the Ancon Hospital grounds, wrote that landscaping might be a non-coercive way to enroll residents in sanitation efforts: "I believe it is a vast help to the Sanitary Department if it takes upon itself the task of propagating systematically such [ornamental] plants, and distributing them among people that apply for them. There will probably spring up a rivalry between people as to who can make his place look best; weeds and tall grass will disappear from the vicinity of the dwelling houses, and surroundings will be improved as to the sanitary condition."[21]

In 1907, Schultz was transferred to the Department of Labor, Quarters, and Subsistence—the department charged with feeding and housing the workforce. Schultz's new boss, the chief quartermaster, encouraged him to experiment with growing the fresh vegetables desired by and recognizable to North Americans—tomatoes, cucumbers, beans, eggplants, okra, radishes, and lettuce—but the gardens received limited institutional support. Whereas canal administrators saw the "healthy landscapes" associated with manicured lawns, ornamental plants, and sanitation programs as vital to the well-being of white employees from the temperate zone and, therefore, a suitable use of canal appropriations, many saw vegetable gardens as increasingly unnecessary as imported fresh foods became more readily available through new transportation infrastructure and refrigeration technology.[22] Thus, the canal company allowed its gardens to go fallow until they received Bennett and Taylor's scientific recommendations concerning the agricultural possibilities of the Canal Zone.

"In so far as the Canal Zone is concerned," Bennett and Taylor wrote in their 1912 report *The Agricultural Possibilities of the Canal Zone*, "the total present crop production barely supplies the simple needs of the scattered rural population."[23] They identified the region's important agricultural features as a humid tropical climate, broken topography, clayey soil, production dominated by mixed and migratory cropping, a large population of Panamanians and West Indian laborers, and public lands held by the US government that could not be titled. In 1909, three years before the depopulation policy was implemented, agriculture was "confined mainly to the meager efforts of the native and West Indian population and they are restricted to patch farming."[24] For Bennett, the farmers' "primitive" techniques—described as "migratory cropping" on small plots without crop rotation or plowing—were a natural outcome of the bounty and enervating effects of the tropics and an impediment to development.

"The native," he wrote, "is an independent person who is not always ready to work, even for the best of wages, because of the contentment he finds upon his small clearing in the midst of sufficient fruits and vegetables to meet the food requirements of his family, with a small surplus for providing the few additional wants."[25] This racialized stereotype of tropical fecundity and laziness was a key reason why administrators sought to block canal laborers' access to land.

What Bennett and Taylor did not see when they arrived on the isthmus was the smallholder banana export economy that had thrived along the Chagres River and the parallel railroad line in the nineteenth century (chapter 5). Instead, they saw a rural landscape emptied out by another construction boom. The scientists' limited understanding of regional environmental history surely shaped the agricultural production that they described in the report.

"The most promising line of attack upon the agricultural problem of the Canal Zone," the scientists concluded, "will apparently be to develop a permanent mixed tropical agriculture with a distinct horticultural trend, in which hand labor of tropical origin will be the main dependence for tillage. [It should] favor the production of high-priced products requiring regular and frequent transportation service, such as will doubtless be available promptly after the opening of the canal for use."[26] West Indian labor, they wrote, promised to be abundantly available after construction was complete and the number of workers needed thus dramatically reduced. Moreover, Bennett and Taylor saw West Indians as more industrious agriculturalists than "native" Spanish-speaking people.[27] They were optimistic about West Indians' agricultural potential under "competent supervision": "A considerable part of the population of this character could maintain itself on the products of the soil if either encouraged or permitted to remain when the completion of the canal ends the need for labor on construction work."[28] But would unemployed laborers be permitted to farm in the Zone after construction?

Figure 8.4
Agricultural experts saw West Indians as more industrious than "natives." *Source:* Hugh Bennett and William Taylor, *The Agricultural Possibilities of the Canal Zone* (Washington, D.C.: Government Printing Office, 1912), Plate II.

Chief Engineer Goethals read Bennett and Taylor's 1912 report as scientific support for the rural depopulation order implemented that same year (chapter 6). According to Goethals, "natives" were incapable of producing food on a scale large enough to justify agricultural development. Moreover, he saw their presence in the Zone as a liability for his government: "I did not care to see a population of Panamanians or West Indians occupying the land, for these are non-productive, thriftless and indolent. They would congregate in small settlements, and the cost of sanitation and government would be increased materially through police, waterworks, sewers, roads, fire protection, and schools." He wrote that US farmers, by contrast, might be capable of sufficiently large-scale production, but it would be difficult to convince them to work in the Zone because they could not be given legal title to land, due to the terms of the canal treaty with Panama (i.e., the US government itself was leasing much of the Zone from Panama). Moreover, it remained unclear which and how much land would be needed for canal purposes in the future.[29]

Finally, some administrators, including Goethals, and members of the US military argued that a depopulated, roadless forest afforded the canal a natural defense against invading forces. Goethals wrote, "With the jungle large bodies could not be moved with ease and rapidity, and though small parties might work their way to the locks, they could do no damage if the defense was alive to its duties." Ironically, the characteristic traits of tropical nature that the canal builders had struggled against during the construction period—heat, insects, and disease—were reimagined as a green buffer to buttress the waterway's defenses. "All the arguments," he concluded, "seemed to point to the desirability of depopulating the Canal Zone, thereby decreasing the cost of civil government and sanitation and increasing the protection of the canal. This was therefore advocated."[30]

The Canal Zone "Thrown Open" to Agriculture

The Panama Canal opened during the beginning of the First World War. The war precipitated a global recession that intensified the economic shock on the isthmus associated with the completion of construction. After the canal was open, only a small portion of the massive construction-era labor force was still needed. The number of employees peaked at more than fifty-six thousand in 1913.[31] In 1914, canal administrators laid off thirty-eight thousand employees. Over the next three years, some thirteen thousand West Indians accepted free trips home courtesy of the US government, per the terms of their contracts.[32] Others migrated across Central America looking for work, some finding jobs at new company banana plantations. But most Caribbean migrants—an estimated forty to fifty thousand, including former laborers, their families, and others attracted by the construction boom—remained on the isthmus. And, to make matters worse, new migrants continued to arrive from the Caribbean, where the economy was even worse than in Panama.[33]

Rents were high, jobs scarce, and wages low in Panama City and Colón, amplifying economic, racial, and class tensions. As urban populations grew, the rural Canal Zone remained depopulated, as it had been since 1912. West Indian leaders campaigned to have those lands opened to agriculture to mitigate the effects of the economic downturn. In 1917, Reverend S. Moss Loveridge, a black Baptist leader, pleaded with Canal Zone Governor Chester Harding, Goethals's successor, to consider resettlement:

Just now, when in the United States and Europe, so much stress is being laid on the question of intensive cultivation, and when the soldier makes preparation for any possible eventuality it seems to me that—(a) With tens of thousands of acres of uncultivated land on the Isthmus. (b) With thousands of unemployed at the present time in the cities of Colon and Panama, both of the "out-of-work," and "won't work," classes. (c) With the present organization and agricultural experts of the United States Government on the Isthmus. (d) With the planting season just before us, and (e) Cooperation between the Governments of the United States and Panama; That it would be a splendid opportunity to utilize the idle labour and idle land, if the United States would furnish the appropriation to cover same, and if necessary conscript all unemployed labor for the purpose of placing the Isthmus, with its present military and civilian population, a long way towards becoming self-supporting so far as vegetables, fruit, etc., are concerned.[34]

Canal administrators began to consider resettlement in earnest after a massive strike in 1920, during which fifteen to seventeen thousand laborers walked out of work, nearly bringing canal operations to a halt. The 1920 strike, like smaller strikes in 1916 and 1919, was an effort by largely West Indian silver roll workers to protect benefits and pay in the face of cuts and a rising cost of living.[35] Proponents within and outside of the

Zone government argued that an agricultural land lease program would reduce mounting agitation on the isthmus by relocating former canal laborers from crowded, expensive, and tense cities to areas where they could farm for subsistence. And, from a more cynical managerial perspective, an agrarian program could sustain a "labor reservoir" to be tapped for future construction work around the canal, while producing food for employees.[36] In 1921, the US secretary of war appointed a commission to make recommendations on the issue.

The commission's report recommended that the Canal Zone be reopened to agriculture. Like his predecessors, Jay Morrow, the Zone's third governor, expressed dissatisfaction with the idea of allowing West Indians—whom he considered ignorant, lazy, and myopic—to repopulate rural areas. "These people have no money to put up proper houses or to do any proper cultivating," he wrote, "and I would be reluctant to see the Canal Zone again covered with the unsightly and unsanitary shacks which once existed in the neighborhood of all Canal Zone towns."[37] Nevertheless, he received instructions from the secretary of war in October 1921 to implement a land lease program that would "throw open" depopulated rural lands to agriculture. The new land leases would be restricted to former canal laborers, excluding most Panamanians. The restriction reflected the fact that the key objective of resettlement was not agricultural development, per se, but managing the frustrated and potentially unruly former construction labor force in the terminus cities.

The land lease program operated according to a different rationality than agricultural development in the Republic of Panama and other so-called banana republics in Latin America. Under President Belisario Porras, the Panamanian government gave massive land concessions to European and US capitalists in the nation's rural interior in exchange for commitments to build transportation routes and provide jobs. Bananas, which made up more than half of Panama's exports, were central to modernization efforts.[38] In the Canal Zone, by contrast, large-scale private farming was untenable given tensions between the United States and Panama over territorial sovereignty in the enclave. Paradoxically, the valuable agricultural lands around this archetypal modern project would be cultivated almost exclusively by poor West Indian smallholders.

Banana Fever, Again

The terms of the land lease program were excellent for former canal laborers: no rent for the first two and a half years, free treatment in Canal Zone hospitals for seven years, and free home construction materials.[39] As resettlement began, one local newspaper columnist wrote, "The bushman had been allowed to come back to his native habitat."[40] But, in fact, neither the "bushman" nor the habitat was native. First, lessees were overwhelmingly foreign-born: 87 percent of the nearly 900 leases went to West Indians during the first six months of the program.[41] Second, the rural landscapes that the former laborers resettling the Canal Zone saw in 1922 might have been unrecognizable to someone who had lived along the Chagres River two decades before. The floodplains farmed by previous generations of banana producers were beneath the lake (chapters 5 and 6). Former hillsides had become lakeshores and old hilltops had turned into forested islands. Land cover had also changed. Between depopulation in 1912 and resettlement in 1921, secondary forest had grown up across the Zone.[42]

Gatun Lake was created to provide water for global shipping, but it also presented new opportunities for regional production. In an oral history interview, one man whose father was a banana farmer on the lake in the 1920s described the emergence of banana production with the new land lease program as far from surprising. *"Subió la fiebre del banana de nuevo"* (banana fever arose again), he explained to me, almost impatiently. "When the settlers came, they said, 'You can plant bananas here. Here is the lake, and there are forests nearby to cut and replant with bananas, too. They won't have to be carried far, just cut down and thrown in the canoe.'"[43] Although most lessees were new to Panama, they were not new to bananas; many West Indians had worked on banana plantations in the Caribbean and Central America.[44] Here, like along railroads during the previous century, their agricultural knowledge facilitated the establishment of banana networks around new infrastructure.

Panama Canal bananas were transported across regional aquatic networks of canoes, launches, lighters, and the railroad—which ran over a raised causeway that split Gatun Lake—before reaching the port in Colón, where they were loaded on oceangoing steamships. The banana linked white and black people, English and Spanish speakers, smallholders and capitalists, machine and machete, and tropical and temperate zones. "Everybody was planting, and everybody expected to make from 50 to 100 per cent on the investment," a journalist wrote. "More than one prospectus was issued describing bananas as 'green gold.'"[45]

Figure 8.5
During the 1920s, bananas were called "green gold" on Gatun Lake. *Source:* Personal collection of Vincente Pasqual.

Chagres River bananas had been global commodities fifty years before and now Canal Zone bananas were also being exported. By 1925, nearly all of the Zone land with easy river or lake access eligible for land lease had been planted with bananas.[46] Reviving a practice that dated to riverside banana markets in *pueblos perdidos* like old Gatun decades before, new lakeside communities like Limón held weekly market days. On those days, smallholders with farms nearby would cut enough bunches—often fifteen to twenty—to sell for the cash necessary for a week, while leaving enough bunches to sell the following week. Farmers carried the bunches on their backs to canoes, with the bananas wrapped in leaves so they would not be bruised. Buyers might arrive in towns on Gatun Lake to find a hundred canoes gathered with fruit to sell. These were prosperous times, with workers making fifty to seventy-five dollars weekly (roughly seven hundred to one thousand dollars in 2014 buying power). Bands of musicians traveled around the lake, playing a night or two in each town. Farmers had cash to buy goods in stores, but still planted subsistence plots of corn, rice, plantains, and root crops alongside their small banana plantations.[47]

The phrase "green gold" was appropriate in a dual sense: banana plants generated vast, often ephemeral, wealth and green bunches of fruit retraced the land and water routes across Panama traveled by precious metals from the Andes and California in past centuries. But, for the first time since the demise of the local transportation economy in the mid-nineteenth century due to the railroad, rural people exerted control over an economy along the river, albeit a minor one.

"Green gold" did not apply to all bananas. It referred to standardized fruit of the *Gros Michel* variety that could be successfully shipped to and sold in North America. Export standards shaped how people interacted with bananas, one another, and the world. Picked green in the tropics and sold yellow to consumers in temperate climates, each Canal Zone banana bunch had to be unbruised and seven to ten "hands" high (rows or tiers) to qualify for export, with each hand composed of fifteen to twenty "fingers" (individual fruit). Banana standards structured negotiations between farmers and buyers: it was not uncommon for farmers to be unable to sell "sub-standard" fruit they had cut, carried, and canoed to market. In practice, standards were surprisingly malleable and political. A 1935 description of the aptly named Standard Fruit's buying practices resonates with oral histories of Canal Zone banana negotiations.

No matter how conscientious an individual inspector may be, it is hard for him not to be influenced, consciously or unconsciously, by various psychological factors in drawing the fine distinction between good and bad quality or between one grade of maturity and another. When the company is clamoring for more fruit, or when a planter influential in politics is concerned, the inspector tends to be more lenient than usual in inspecting the fruit. Sometimes, however, there is more available fruit than the steamer at the dock can carry, or market prices are dropping and a sizzling radio message arrives from Boston, saying, "Reduce your cargoes," or "Ship only good fruit." Then the tropical division manager and his superintendent of export, who do not want to be blamed for payments for large quantities of fruit that the company prefers *not to accept and market*, check up on the inspectors, who often work 18 to 24 hours at a stretch, ordering them to be especially careful not to receive any but "the best fruit."[48]

The Zone's land lease program provided laborers-turned-farmers with prime canal frontage plots and access to the railroad and nearby ports. Smallholders dominated production and banana capital was embodied by the *compradores*—middleman buyers—who drove boats around the lake, pulling flat-bottomed barges to carry fruit. Typically independent contractors, *compradores* mediated the economic relationships between the banana export companies and smallholders. For example, they provided loans to new farmers to hire laborers, clear the forest, and plant bulblike corms (a few cents each).

One man, known as the "Banana King of Panama," exerted great power over banana networks on Gatun Lake in the early 1920s. John "Johnnie" Walker arrived in Panama from the United States around 1907, the beginning of canal construction, intending to explore the country's commercial possibilities. He first worked in the lumber trade, selling mahogany and other trees harvested in Panama in the Canal Zone. Through this work, Walker claimed, he came to understand the "natives of the little jungle towns ... their problems, their life, and their feeling."[49]

Walker was not a lessee himself (he was ineligible because he was not a former employee), but he was the embodiment of the white American managerial type that Bennett and Taylor believed could transform Panama's abundant labor, fertile soil, and plentiful rain into agricultural profit. In a 1925 newspaper profile, he was credited with establishing the regional banana industry "on a successful, business-like status ... justifying the hopes reposed in soil and climate." He owned and operated a plantation in Panama just outside the Canal Zone with access to the western shore of Gatun Lake. He paid his laborers by "piece work or hourly, in some cases, for no one but a theorist would ever think of paying a Central American native or West Indian help by any other contract." The natives, he asserted, have little use for paper money, but were "buying $15.00 shoes and tailored clothes" after Walker "taught" them how to grow bananas in quantities to sell.[50] He also made loans to smallholders to cover input costs.

By operating a plantation and trading posts in nearby villages, Walker embodied the strain of intimate, paternalistic capitalism that briefly defined the banana trade around the canal, but changed as large companies attempted to exert control in the region.[51]

In 1921, when the land lease program began, only one company—The Panama Railroad Steamship Line—shipped regional bananas from the Colón port. Unlike many banana exporters in Central America, the Panama Railroad provided transportation and was not involved in production. Regional exports ranged from eight to nine thousand bunches monthly.[52] At the time, the only mature plantations in the region were located beyond the Canal Zone boundary in Panama. By April 1922, the United Fruit Company had begun weekly shipments of bananas from Colón to New York. A year later, they were shipping twice per week.[53] United Fruit was not only the world's largest banana company, but its largest agricultural enterprise. At that time, the company purchased about 60 percent of its fruit from "smaller" planters—a range of smallholders and foreign and domestic capitalists. It was transitioning from buying from private planters, like those on Gatun Lake, to company plantations.[54]

Figure 8.6
Bananas for export at the Panama Railroad pier at the port in Colón. *Source:* Personal collection of Doug Allen.

Johnnie Walker was United Fruit's main buyer around the Panama Canal. In 1924, he traded twelve to fifteen hundred bunches weekly.[55] Both banana companies and farmers depended on *compradores* like Walker to serve as intermediaries, but United Fruit did not like the power that he exercised in lake communities. In a letter to company headquarters in Bocas del Toro, Panama, the United Fruit manager D. O. Phillips concluded, "... it is out of the question for Mr. Walker to control the production in Gatun Lake Region."[56] Phillips suggested that the company buy directly from farmers, cutting out Walker and the *compradores*.

Attention to banana networks on Gatun Lake reveals how an environment created and managed primarily for navigation purposes was reworked from its margins to serve different purposes and communities. By the end of 1922, the land lease program's first year, the Panama Railroad did not have enough dedicated banana cars to handle the volume of fruit produced around the lake for export.[57] In October 1923, perhaps as a response to this bottleneck, small boats carrying bananas from the lake began to transit the Gatun Locks for the first time.[58] Meanwhile, the Panamanian government gave large land concessions around the borders of the Zone to capitalists who promised to plant bananas, build roads, and employ Panamanian laborers. Two new companies—the San Blas Development Corporation (owned by United Fruit competitor Standard Fruit) and the American Banana Company—entered the fray. Spanish-speaking farmers from the province of Coclé, to the west of the Zone, began to migrate seasonally to work in the lake region as banana laborers.[59] Smallholders cultivated forested, sparsely populated lands located farther up the Chagres River and its tributaries. The number of bananas exported from the region doubled almost every year between 1922 and 1925.[60] By 1925, the Panama Railroad

Table 8.1
Banana Exports from Canal Zone Ports

Year	Number of bunches
1922	208,688
1923	309,716
1924	840,321
1925	1,727,491
1926	2,182,688
1927	2,773,792

had dedicated two to three full trains weekly to transporting bananas from Gatun Lake.[61]

That same year, United Fruit brought the first steamer to buy bananas on Gatun Lake—a practice that administrators had previously forbidden on Panama Canal waters. The lake, after all, was created to serve as a water storage reservoir for shipping, not farming. The steamer cut the *compradores* out of the network.[62] The company sought to consolidate its control of production on the lake by bringing steamships onto the lake and locking smallholders into contracts.[63]

F. oxysporum as an Infrastructural Species

The same global transportation infrastructure that facilitated the emergence and expansion of the banana plantation model eventually threatened its success. Early in 1926, Panamanian newspapers announced that banana blight, which had decimated United Fruit plantations in Bocas del Toro on the country's Atlantic coast was spreading rapidly around Gatun

Figure 8.7
Banana rhizome, or corm, with emerging "sucker." Photo by the author.

Lake. The *Panama American* saw dark clouds on the horizon: "There has been no falling off as yet in total production, and there may be none for some time to come, but the blight is spreading rapidly, and various planters who expected large returns on their investments will now be satisfied if they ultimately recover what money they expended."[64]

The first visible symptom that a banana plant had been infected by *Fusarium oxysporum*—the fungal pathogen that caused Panama disease, which was known as "banana blight" around Gatun Lake—was the yellowing of the oldest leaves. Then, younger leaves wilted until the plant was covered in dying leaves. The infection did not immediately kill the plant, but decreased the quantity and quality of the fruit produced. By the late 1920s, the blight had devastated plantations in Bocas del Toro, Panama to the point that production ceased altogether.[65]

F. oxysporum was an infrastructural species, like the fruit it traveled with. Trains, steamships, and the monoculture plantation model constituted both infrastructure for banana production and invasive network for Panama disease, because they carried infected banana corms, or rhizomes. The soil-borne fungus remained isolated—fixed in place—as long as plants and plantations were not directly next to one another. However, the transformation of forests and multicrop farms into banana plantations enabled the fungus to spread more easily within localities and regions. Many biological connections, however, were established through long technical networks rather than geographical proximity. The global infrastructure assembled by banana companies enabled *F. oxysporum* to leap across national borders and between ecosystems, because when companies established large plantations in new areas, they often imported hundreds of thousands of potentially infected rhizomes from other countries.[66]

The troublesome fungus was probably not new to Panama, then, but the territorial expansion of the banana as an infrastructural species also facilitated its success. As monoculture production came to dominate more of the Chagres River landscape, soils became more biologically interwoven through the transfer of corms used in propagation among farms. Their flow through the Canal Zone is difficult to retrace, but a 1926 advertisement in the *Panama Times* offering twenty-five hundred hectares of "Virgin Banana Land" in Panama near Gatun Lake provides a clue. For the purchaser, the ad stated, corms would be "cheaply and conveniently obtained" locally.[67] One local producer saw rhizome transfers as the key problem, noting that "unscrupulous parties operating on the Lake sold suckers from old diseased plantations to unsuspecting planters, and widely spread the blight."[68]

The Decline

The quantity of bananas shipped from Canal Zone ports peaked just under 2.8 million bunches in 1927.[69] This was the high water mark. There is no easy answer as to why banana networks around Gatun Lake fell apart. The expansion of *F. oxysporum* certainly played a role, but it alone did not cause the decline. There was no massive influx of wage labor opportunities, as there had been in the region with the initiation of the French canal project in 1881 and US canal project in 1904. Ecological, economic, and governmental factors contributed to the decline.

United Fruit moved to consolidate its economic domination of the banana trade on the lake. Tensions between the company and producers came to a head in June 1927 when weekly purchases were unexpectedly cut by half—from thirty-two to fifteen thousand bunches—and the company announced that it would only buy from farmers under contract going forward.[70] Devastated, independent producers accused the company of using its clout to force them into unfair contracts. United Fruit responded that it no longer had the capacity to buy low quality bananas from the canal area as production increased at its company-run plantations. More problems emerged as the 1930s began. The onset of the global economic depression led to decreased demand for bananas in US markets. Prices fell to less than half of their mid-1920s levels.

The Canal Zone land lease program restricted the size of banana farms to fifty hectares, but the average land area leased was around four hectares per licensee.[71] This type of smallholder production, which was becoming outdated in the 1920s due to the expansion of plantation production, mimicked the decentralized business model that historian John Soluri calls "vernacular agriculture." In 1931, Canal Zone land agents reported that farmers were unable to pay their leases due to difficulty selling fruit: "Most of our licensees are ex-canal employees, advanced in years, and very poor, and the collections are more difficult to make than formerly, the greater part of the rentals being collected by the land inspectors only after repeated calls on licensees in inaccessible locations."[72] Many farmers lived on the banks or islands of Gatun Lake, so the land inspectors used a canoe to collect rents, often with great difficulty.[73]

Five short years after the Canal Zone was thrown open, canal land agent N. A. Becker reported "exhausted soil" across 600 hectares, nearly 10 percent of the leased land in production. This might have been an unrecognized effect of *F. oxysporum*, as the symptoms of infection—leaf wilt and decreased fruit production—were similar to those of soil deficiencies.[74] He recommended that the 257 lessees in the denuded region (mostly "aging West Indians") be allowed to move elsewhere in the Zone. Gatun Lake banana production did not decrease precipitously, but growth leveled out and production was geographically redistributed as smallholders abandoned unproductive farms near the lake for forested "virgin" lands higher in the Chagres River basin.[75]

Conclusion: The Forest Returns

The questions of *how* and *by whom* rural lands in the Canal Zone should be used were difficult to resolve. The depopulation policy was effectively reinstated in 1932 when the land lease program was discontinued after years of debate over the costs and benefits of allowing smallholders to occupy rural lands. While some canal administrators saw the black West Indian farmers participating in the program as a "labor reservoir" to be maintained for future projects, others saw them as a public health menace to white Zone communities or a drain on state services like hospitals.[76] The racial biopolitics of sanitation still haunted the Canal Zone (chapter 6). The governor of the Canal Zone described the decision to depopulate rural areas again in his 1932 annual report: "It is not practicable to care for any number of these people by allowing them to settle on land in the Canal Zone; many could not make a living in the jungle and the increases in malarial infection which have resulted in the canal towns from the presence of settlers on the land have led to the decision to license no more settlers."[77]

In a letter to a colleague in 1930, Canal Zone horticulturalist J. E. Higgins explained that migratory banana agriculture around the canal was not simply a case of backward resistance to modern agricultural methods, but a rational response to the dynamic environmental

Table 8.2
Land Leases and Population in the Canal Zone (1920–1932)

Year (as of June 30)	Number of leases	Population
1920	—	10
1921	—	21
1922	1026	203
1923	1805	810
1924	2154	2387
1925	2112	2523
1926	2012	2684
1927	2019	3395
1928	2041	3568
1929	2123	4597
1930	2102	4482
1931	1927	4129
1932	1785	3719

Source: N. A. Becker to Colonel J. L. Schley, June 30, 1930, NACP RG 185, Entry 34 File 33.b.51(7); and J. F. Siler, Chief Health Officer, to Committee to Reopen the Canal Zone to Settlers, February 2, 1933, NACP RG 185, Entry 34, File 34.33.b.51(8).

conditions created through the articulation of the banana, transportation infrastructure, markets, and governance. Higgins argued that smallholders around Gatun Lake were locked into a cycle of deforestation, cultivation, and migration to produce the standardized Gros Michel banana demanded by North American consumers—a variety highly susceptible to infection by Panama disease. These farmers, he wrote, failed to invest intensive labor and capital in the land because they knew they would farm a plot for only a few years before it was decimated by the blight. Panama disease diminished production, rendering producers' bananas less attractive to buyers.

The banana illuminates the contested history of land use around the Panama Canal. As we've seen, agriculture was not peripheral to the history of transportation on the isthmus, but bound up with it (chapter 5). Even as coalitions of administrators and labor leaders pushed for rural settlement and regional agricultural development, others actively opposed the plan. The Canal Zone land lease program that began in 1922 and ended a decade later gave rise to highly charged inter-institutional tensions. Communities thrived along the margins of the waterways that white administrators could not completely regulate or control, a marked contrast to the organized grids of Zone communities constructed around the canal termini and next to the Panamanian cities of Panama City and Colón.

Consider, again, Canal Zone horticulturalist O. W. Barrett's provocation in the epigraph, written in 1915: "In three hundred twenty-five square miles of moderately suitable territory, more or less isolated, how nearly self-supporting shall this area be made, what sort of horticultural development should it have and how much?" Looking back from the twenty-first century, the answer seems clear: transportation and the service economy have displaced agriculture and rural production. Through this shadow history of the Canal Zone, I hope to demonstrate that, though it was far from perfect, an alternative vision of the relationship between transportation, agriculture, and the regional economy did exist. It makes one wonder what a more diverse, self-supporting economy might look like today.

Part III   The Interior

9 Getting Across and Getting Around

Figure 9.1
Watching ships from the side of the highway, Pedro Miguel Locks, 2009. Photo by the author.

While conducting research, I often drove between Panama City and my field sites in the rural areas around the canal. To get to Limón or Boquerón, you drive north from the city on Avenida Omar Torrijos Herrera (figure 9.2). Formerly known as the Gaillard Highway, the road runs along the east bank of the canal. Beginning near the Panama Canal administration building in Balboa, it skirts the former Canal Zone community of Albrook, now a suburb with shopping centers, chain fast food restaurants, and manicured lawns. Past Albrook is the passenger station for the Panama Railroad, which is still in operation 150 years after its construction. A "landbridge" complement to the Panama Canal, the railroad primarily carries shipping containers today. Its passengers are not forty-niners bound for Californian gold fields, but Panamanian businessmen commuting to Colón's free trade zone and tourists in search of canal views. Driving on the highway, you pass by power plants, machine shops, administrative offices, and dry docks—all infrastructure for interoceanic transport.

Next, you pass Fort Clayton, the hulking former US military base reimagined as the City of Knowledge. It houses organizations like the United Nations Development Program and Nature Conservancy rather than soldiers. Then, you see the Miraflores Locks Visitor Center, where foreign tourists arrive by the busload and pay to see how the canal works. By contrast, the Pedro Miguel Locks—only a few miles farther north up the highway—are frequented by Panamanian families and couples, who watch ships in transit through a chain-link fence, free of charge (figure 9.1). Beyond Pedro Miguel, one-story pink and gold houses line the road as it winds through Paraiso, a community built for West Indian and other nonwhite canal laborers. Not far past there, a green lawn dotted with small white crosses slopes up from the side of the road, a reminder of the violence of construction on the isthmus. The cemetery memorializes a small number of the estimated twenty thousand lives lost during the French canal project.[1]

The road splits and—if you're heading to Limón or Boquerón—you veer right (east) on Madden Road, another old Canal Zone road, and travel away from the waterway. The canopy swallows the sky as you enter a forest that feels ancient and insulated from the world—until a honking bus comes around a bend in the road. If you're paying attention, you might notice a quieter, older artifact of global connection: a sign marking the colonial Spanish road, Camino de Cruces.

You cross the former Canal Zone-Panama boundary some five miles later. The forest thins and cracks radiate across the pavement. You turn left onto the Transístmica (transisthmian highway), headed north again. The transition from the former Canal Zone, where the Panama Canal Authority

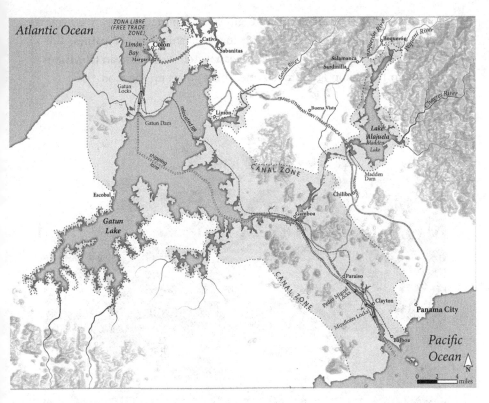

Figure 9.2
Land and water routes across and around the transit zone. Image by Tim Stallman, used with permission.

has maintained much of the uniform North American architecture and landscape aesthetic, is jarring. The highway is a world of concrete: the hum of rubber on pavement, the odor of acrid exhaust fumes, the visual jumble of cinder block buildings and weedy lots.

The Transístmica crosses the Chagres River below Alajuela Lake (formerly Madden Lake). *Diablo rojos* (red devils), the brightly painted former US school buses that have enjoyed lengthy second acts in Panama, are everywhere.[2] They roar and belch black smoke, lurching in and out of traffic to reach crowded bus stops. The roadsides are lined with grocery stores that sell hot dogs and colorful sodas. Many of the stores are owned and operated by Chinese-Panamanians—the customers typically call them *chinos*, both the shops and owners. There are also signs that the highway is, like the canal, linked to global transportation networks. Tractor-trailers pull containers with the logos of shipping companies. Billboards promote American jeans, European watches, and Korean televisions for sale in the Colón Free Trade Zone, thirty miles up the road.

Figure 9.3
Diablo rojos are privately owned former US school buses that serve as de facto public transportation in many parts of Panama. Photo by the author.

The highway corridor is a world in motion, but it is also a chokepoint.[3] In the event of construction work or an accident, traffic comes to a halt. When I conducted Colón Free Trade Zone, field research in 2008–2009, it was the only highway that ran from ocean to ocean. There were no alternative routes. Although a parallel toll road has since opened, it is an expensive expressway between Panama City and Colón that bypasses most of the in-between communities along the old highway.

Turn left, or west, on any of the roads past the Chagres River and you descend toward Gatun Lake. One of them goes to the town of Limón. Turn right, or east, and you drive toward rolling green mountains and into the river's headwaters. If you are driving to Boquerón, you take a right at the second major intersection past the river. Off the highway, the scenery becomes rural. You weave among walking groups of schoolboys in white polo shirts and green slacks chatting up schoolgirls in matching skirts. Or a small red pick-up truck heavy with fruit creeps ahead of you, the driver droning an incantation of *"plátano, guineo, mango, papaya"* through his rooftop megaphone. A campesino man in a straw hat, pants tucked into knee-high rubber boots, walks with a machete. Barbed wire fences demarcate teak forest plantations and cow pastures. A dog sunbathing in the road struggles to its feet and moves, begrudgingly, to allow you to pass.

This rural landscape is less than thirty miles by road from Panama City and the Miraflores Locks, but it is a world apart. The people who travel this road regularly will tell you that it is *feo* (ugly). The surface is part asphalt, part gravel, and almost all in need of maintenance. Passengers standing in the crowded beds of taxi pick-up trucks sway as their drivers carve wide S-patterns to avoid abusive ruts and washboards. Up the road further, the power lines disappear and the topography buckles. The road, all gravel now, climbs and winds through small mountains and the houses become fewer and recede further from the road. The Boquerón River appears on your right as you continue to climb.

You pass a sign and an unmanned guard post that marks your entry into the Chagres National Park and signifies that these forests are protected parts of the Panama Canal watershed. Your car is the only one on the road now. Clay-red puddles fill the depressions that fleck the road, the remains of a downpour a few days earlier. You swerve to avoid them. The crunch of your tires on gravel announces your arrival in Boquerón. The families along the road know you are here before you do. Faces peek down from the straggle of roadside *ranchos* (open-sided structures with palm thatch or tin roofs) and cinder block homes to see who you are. They shout greetings if they know you and stare if they don't.

Boquerón is a small community that is getting smaller.[4] Since the 1980s, many families have moved to places where life is a little easier, working in the urban service sector or farming in areas like the Darién province, where agriculture is less restricted than in the Panama Canal watershed.

Community members are always coming from or going to the Transístmica. They leave town, or "go outside" (pa' fuera), to work as laborers, sell agricultural products, shop at discount megastores, seek government services, and visit friends and relatives. Getting around is not easy here. There are only two private cars in town, so nearly everybody depends on the bus that rumbles up the gravel road four times per day, assuming that it is running and the road is passable. Boquerón is the last bus stop, so those who live farther up the road continue by foot or horseback—or, if they're lucky, catch a ride with a passing truck.

The bus links life here with the rhythms of the highway. Due to local dependence on the bus and the road, connections with the "outside" are precarious. For example, a strike by the transportistas—a national union of bus, taxi, and truck drivers—brings public transportation to a halt, leaving people in Boquerón and other communities facing hours of walking. These difficulties signal both the infrastructure that has never arrived and that which has come and gone.

When canal authorities enacted conservation policies in the area in the 1980s (chapter 3), they encountered a landscape sedimented with infrastructure projects. The nearby ruins of the Camino Real were centuries old, but there were also more recent lines through the forest. In 1915, a North American businessman living in Colón tipped off by a rubber tapper found manganese near the river and built a narrow-gauge railroad over the mountains to the Atlantic Ocean. Then, manganese prices fell and, by the 1940s, the railroad had been abandoned. In the 1930s, the US canal company built the hydrographic station across the river from the current site of the community to collect data for water management downstream.

The most recently abandoned infrastructure here is that built by the Panamanian state in the 1960s and 1970s to encourage rural colonization, development, and agricultural modernization. Due to shifting political and economic priorities, it has been retracted. The material and cultural

Figure 9.4
The road to Boquerón, 2009. Photo by the author.

legacies of this era point to a historical dimension of environmental problems around the canal that is often elided. Campesino farmers were not representatives of a backward local culture spontaneously destroying forests, but participants in a multiscale development project that was designed to transform rural areas by extending infrastructure and expertise.

When Panama Canal administrators sought to protect watershed forests as part of a "global" transportation infrastructure, then, the existing "global" project that had already territorialized the same area became a problem because the projects were in social, economic, and environmental conflict. As one infrastructure was extended and the other retracted, circulation downstream by water was protected, while circulation upstream by land became more difficult. The next two chapters retrace the extension and retraction of rural development infrastructure and analyze how that historical process has articulated with everyday life.

Figure 10.1
Campesinos navigate a dirt road near the Transístmica, 1962. *Source:* Smithsonian
Institution Archives, RU 7006, Box 187, Panama 1962. Used with permission.

The Zone is like Egypt; whoever moves must travel by the same route.
—Harry Franck, travel writer and Canal Zone policeman, 1913[1]

The Panama Canal operates smoothly for the most part, but getting around the surrounding region takes an inordinate amount of residents' time, money, and energy. From rural villages to urban centers, friction—not flow—is the dominant experience of movement: bouncing on bus seats, standing and swaying in crowded pick-up beds, and sitting in traffic jams. The juxtaposition of global flow and regional friction in Panama's transit zone runs deeper than irony, for both are outcomes of how networked infrastructures are built, managed, and maintained.

On the one hand, roads, railroads, and engineered waterways facilitate connections between communities. On the other hand, transportation infrastructures disconnect communities, whether by design (which sites they do and do not link), through governance (who is allowed to access them), or by establishing physical barriers that restrict use of older routes.[2] Infrastructures channel people and things along specific routes in support of social, economic, and political projects that reflect the priorities and values of their builders. When global infrastructures and regional-to-national networks occupy the same landscapes, they develop in relation to one another, converging and diverging in their purposes and politics.

The story of Panama's Transístmica (transisthmian highway) is a prime example of how infrastructures built to serve different communities and social visions are entangled, producing experiences of connection and disconnection. The construction of this fifty-mile strip of potholed pavement parallel to the canal's shipping lane was funded through United States defense appropriations during the Second World War. When it opened to the public in 1943, it was, shockingly, the first interoceanic road across Panama—the so-called crossroads of the world—since the disintegration of the Camino Real in the nineteenth century.

The Canal Zone's motto was "Panama divided, the world united." Yet, by emphasizing global maritime connection (the world united) through canal construction, we often forget that territorial disintegration (Panama divided) was an outcome of the same process. As the author of a 1925 newspaper editorial calling for a bridge over the canal wrote, "The union of two oceans necessarily brought about a complete and effective separation of the land. The west half of the Republic of Panama is almost as

effectively cut off from the eastern half as if they were two distinct continents. Communication between the two, both commercially and from a
military standpoint, is a slow and difficult matter, necessitating the use of
barges and tugs. No rapid and direct communication is possible."[3]

The opening of the canal reoriented transportation in the surrounding
region and the Republic of Panama. By damming the Chagres River and
flooding Gatun Lake, canal builders drowned river communities—the
pueblos perdidos—and "obliterated" old foot trails and water routes, while
opening up vast new areas to exploration and settlement by canoe and
boat.[4] In the absence of a bridge across the canal, it was impossible to
travel by land from the rural interior to the cities and markets of the
transit zone. Within the context of a disconnected transit zone and an
unconnected rural interior cut off by a foreign-run waterway, national
integration through road construction became a priority for the Panamanian state.

Releasing the Energies of Rural Panama Through Roads

The automobile was less than two decades old when the Panama Canal
opened in 1914. Around the world, paved roads were largely restricted to
urban areas. Beyond city limits, mobility was tied to railroad lines, waterways, and dirt roads. At that time, for example, more than 90 percent of
the roads in the United States were dirt and difficult to drive by automobile; they were rutted when dry and muddy when wet.[5]

In early-twentieth-century Panama, transportation between regions was
restricted to steamship lines, because the roads—even important ones—
linked port cities and towns to inland villages. The residents of Panama
City, Colón, and the surrounding area could travel by railroad and, after
1931, take a ferry across the canal, but there was no rapid interoceanic
route—land or water—beyond the Canal Zone.[6] Meanwhile, in the rural
interior, the dirt roads called caminos de verano (summer roads) were passable by foot and pack animal and, perhaps, by wagon during dry periods.
The Panamanian state had a small bureaucratic and infrastructural presence in the interior at that time. Therefore, the roads that existed were
locally administered routes constructed and maintained by corvée labor.
In lieu of taxes, men over eighteen years of age—only poor men in practice—were required to work on the roads three days every year.

The author of a 1908 editorial entitled "Wanted—Good Roads" in the *Star and Herald,* a Panamanian newspaper, characterized the lack of modern infrastructure as an obstacle to development and argued that "good" roads turned nature into resources:

It is a lamentable fact that the country is still woefully deficient in ways of communication. Between the capital city and the chief port on the Atlantic side the only means of communication is by railway, and that is not owned or controlled by either the Panamanian Government or people. There is no highway between the capital and the chief towns of the provinces; no roads leading from the capital to any important place in the interior in fact. The only communication with other points in the national territory is by water, and that of course is limited to points on the coast or on navigable rivers. The result of this absence of highways is not only to prevent free communication between important places in the country but to hamper its development considerably. The district for miles around Panama [City] should be alive with industry. Instead of which a few minutes' stroll from the city limits brings one to jungle or swamp where mongoose and the iguana pursue their livelihood undisturbed. ... Without roads the resources of the most fertile area must remain untapped or be brought to market at a cost which renders operation a doubtful gain. How can the vegetable and mineral resources of this country be developed to any great extent if the areas in which they are located remain shrouded in impenetrable forest?"[7]

During the presidency of Belisario Porras, whose Liberal administration dominated Panamanian politics between 1912 and 1924, the state pursued an ambitious public works program to build "good" roads and modern institutions linking the rural interior to Panama City, Colón, and the Canal Zone, but they were also concerned that a disproportionate focus on transportation would tie the domestic economy too tightly to fluctuations in world trade, as it had in previous centuries. Thus, the state sought to convert small-scale agriculture into large-scale agricultural commodity production and to pursue integrated rural-urban development. This required changing both landscapes and people.

The government of Panama established the Junta Central de Caminos (central roads department) in 1920 to build a national road network. Its initial mandate was to develop "penetration roads"—dirt roads that would integrate isolated rural areas by facilitating the exploitation of natural resources and stimulating export agriculture and cattle ranching.[8] This language reflected an international discourse about technology's magical capacity to "release the energies" of the land.[9] But roads could not release

Figure 10.2
At work in the Junta Central de Caminos, Panama's roads agency, 1930. *Source:* Government of Panama, *Memoria Despacho de Agricultura y Obras Publicas* (1930).

those energies alone. A modern country also required institutions and scientific experts to cultivate farmers.

Unlike the Zone government, which approached agriculture as a small-scale relief measure and potential source of fresh food subordinate to navigation (chapter 8), the Panamanian state approached transportation networks, particularly roads, as a precondition for modern agriculture. What both governments shared was a disregard for "traditional" agriculture as practiced by campesinos.

Panama established its first agricultural school as part of a larger effort to develop the interior—in tandem with roads and other public works projects—but it was met with indifference by rural people. In 1916, President Porras reported to the National Assembly, "Of 30 stipends available for study in the school, only 15 ... were taken, and even then those without any enthusiasm. The country seems unwilling to awake to the necessity of working the land in order that we may gain economic independence, and is still misled by the illusion that our [geographical] position and abundant natural resources will be sufficient."[10] The administration's idea was that, by attending the school, campesinos would adopt modern techniques like plow agriculture and leave *roza* agriculture behind. These technical and cultural efforts—discussed in more detail in chapter 11—marked the beginning of decades of efforts to modernize rural livelihoods and landscapes together.

Yet, we should not think of transportation in the Canal Zone and agriculture in the Panamanian interior as distinct, because they developed in tandem in the twentieth century. In particular, US-Panama treaties in 1936 and 1955 significantly expanded the market for Panamanian beef, leading to the rapid expansion of cattle production in the interior and, by extension, deforestation.[11]

Foreign Control of Interoceanic Routes in Panama

In Panama and elsewhere, roads were held up as material manifestations of state power, sovereignty, and modernity. For those without them, like rural Panamanians, they were aspirational infrastructure. In 1925, *The Workman*, Panama's West Indian newspaper, visualized national progress in this way:

> The roads leading out from the cities and towns to various sectors of the interior should be pushed to completion as soon as possible. Who cannot visualize the human and bustle of business throughout the whole of this country? With sawmills transforming uninhabited jungle into busy villages; wasting streams lending their liquid vigour to artificial irrigation and all forms of commercial life sent speeding to centres of trade, on rolling freight cars, tooting motor trucks and cutting coasters, while the lonely, snail-like ox-cart, the crawling pack-horse, and the sleepy *cayuca* [canoe], all slink irretrievably into deserving oblivion."[12]

As boosters imagined a technological countryside, Panama's urban merchant class clamored for the construction of a paved highway between Panama City and Colón—a domestic route that would symbolically and geographically parallel the foreign-controlled canal and railroad. Panamanian newspaper editors, who agreed on little, argued that a highway would facilitate free passage—rapid and independent of Yankee control—between the port cities and open the arable land of the Chagres River basin to settlement and agriculture.

Not just any paved highway would do, either. For Panamanian elites and capitalists, modernity traveled at high speed on concrete roads.[13] A 1923 editorial by the Panama Rotary Club, representing the business community, made the case for paving with concrete: "There seems to exist in the minds of a great many a mistaken idea that concrete construction is a luxury to be used only for city streets. This 'hooey' has always been fostered by the advocates of an old semi-permanent type of pavement. Gravel and water-bound macadam might be cheaper in the short term, but require much higher maintenance rates. It is practically impossible to maintain macadam in the tropics."[14]

Building the highway was not simply an engineering problem—a debate over road surfaces and grades—it was also a geopolitical problem. Legally, Panama's roads department could not build a transisthmian highway without the permission of the United States government, even across sovereign territory. The restriction was an artifact of the history of transportation on the isthmus. An 1867 agreement between the Panama Railroad and the Colombian state, which governed the isthmus at the time, granted the company a ninety-nine-year monopoly on "any class of carriage roads whatever, from one ocean to another."[15] When the United States purchased the railroad in 1903, it inherited that concession.

The Territorial Politics of Roads and Forests

The Panamanian state wanted to build roads, level forests, and accelerate the circulation of resources, capital, and people across the countryside. For US government officials in the Canal Zone and Washington, D.C., Panamanian development efforts near the canal raised territorial concerns. As infrastructure, the lock canal was built to facilitate "good" forms of circulation between the oceans (of ships, information, commodities, travelers, and military might). Meanwhile, the Canal Zone was governed to provide water and labor for transportation and to serve as a protective buffer restricting what administrators considered "bad" forms of circulation nearby (of mosquitoes, "natives," and laborers).

Some Zone officials argued that Panama should not be permitted to build the highway. It was not that they saw it as a potential commercial rival to the canal—the original intention of the clause. Rather, they were concerned about public health and canal defense. Whereas Panama wanted to cut its forests to facilitate movement, Chief Engineer George Goethals and other US administrators saw forests as protecting the canal and its enclave by slowing the hypothetical movement of enemy forces around the waterway. But even more worrisome, as discussed in chapters 6 and 8, were nonhuman enemies: disease-carrying mosquitoes. Despite their achievements sanitizing the physical and human geography of the Zone, administrators worried that automobile traffic across a nearby Panamanian highway would facilitate and accelerate disease communication between "native" and white US communities.[16]

Canal Zone institutions were not monolithic. Some administrators were sympathetic to Panamanians' desire to build roads and saw regional development as inevitable. R. K. Morris, the chief quartermaster, argued to the governor in 1921:

The territory outside of the Zone is now settled and will become further settled when these roads are built. With fairly good roads built from these cities [Panama City and Colón] to the interior their means of communication will be improved and the native who is now infected with malaria and only gets to town two or three times a year will be coming in several times a month, and if the Health Department is to

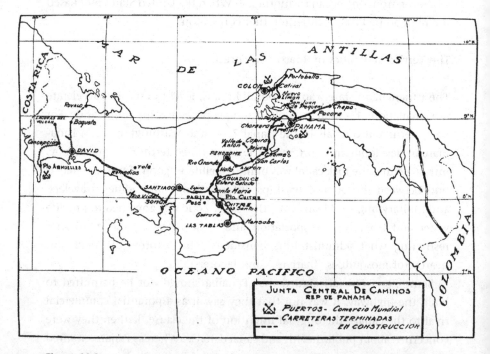

Figure 10.3
Panama's roads in 1930, completed and under construction. *Source:* Government of Panama, *Memoria Despacho de Agricultura y Obras Publicas* (1930).

eliminate infection for this source they will have to greatly extend their sanitary work. ... If sanitation cannot adapt itself to the development of Tropical countries along economic lines then sanitation in the large sense is a failure.[17]

In 1926, the US government agreed in treaty negotiations with Panama to appropriate forty-five thousand dollars to conduct a survey for a trans-isthmian highway.[18] The decision was pragmatic, not an act of generosity. The value of the forest as a military buffer was defined in relation to changing technologies: forest cover might slow the movement of ground forces, but not airplanes.[19] At the same time, the absence of a paved interoceanic road with a network of hard-surfaced auxiliary roads also made it impossible for the United States military to move motorized forces rapidly to the outer periphery of canal defenses.[20] Here, again, government decisions involved weighing the advantages of facilitating "good" circulation against the potential that "bad" people and things would follow the same routes. The political calculus of highway construction was shifting.

The Construction of the Transístmica

After nearly two decades of discussions, the United States and Panamanian governments signed an agreement in 1936 to build a transisthmian highway. The United States agreed to waive the railroad's monopoly rights in order to allow Panama to construct the highway that would come to be known as the Transístmica. For more than eighty years, the Panama Railroad had been the only land route between the nation's largest cities. More incredibly, there had not been a paved road across the isthmus since the deterioration of the Spanish Camino Real two centuries before. By 1936, the Canal Zone government had built a network of highways and roads within the enclave, including the Madden Highway in 1930, which extended to the Madden Dam construction site. With the right to construct the highway granted, Panama planned to connect the end of the existing Canal Zone road system on the Pacific side at Madden Dam to the end of the Zone road system on the Atlantic coast, just outside of Colón.

The Trans-Isthmian Highway Convention was ratified in 1939. As world war loomed, the US military scrutinized Panama's road system and found it underbuilt and poorly maintained.[21] Within a charged geopolitical context, the construction of paved roads, including the long-discussed highway, suddenly became a defense priority. A State Department official wrote, "The reasons for which this Government considers the construction of the Trans-Isthmian Highway to be a project vitally concerned with our national defense are, to my mind, obvious. At the present time the surface means of transit across the Isthmus of Panama are limited to the Canal and a railroad. In these days of motorized and mobile equipment for the armed defense forces of the country, it is of the utmost importance that a hard-surfaced highway be built across the Isthmus."[22]

Therefore, the impetus for beginning the construction of the Transístmica had more to do with a shifting global political and military context than Panama's economic development. The US Public Roads Administration (PRA) was busy building a network of defense roads, both domestically and around overseas outposts. In 1940, President Franklin Roosevelt requested three hundred and twenty-five thousand dollars in emergency defense funds from Congress to build the US portion of the highway over the mangrove swamps between Colón and Cativa (figure 10.4; first section

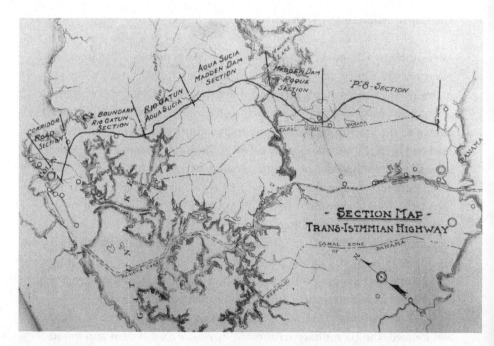

Figure 10.4
Transístmica construction, by section. *Source:* National Archives at College Park, Maryland, RG 59, Entry Decimal File 1940–44, Box 4270.

from left).[23] Meanwhile, the Export-Import Bank of the United States provided Panama with a 3.2 million dollar loan to build the longer stretch of road between Cativa and Madden Dam (figure 10.4; second, third, and fourth sections from left). Under the terms of the loan agreement, the PRA would oversee project expenditures, specifications, design, and construction.

Upon inspection, the PRA found Panamanian plans for their section of the highway inadequate for projected civilian and military traffic. The Panamanian government planned to build a two-lane, eighteen-foot-wide road. But the PRA argued that a wider and more durable road was necessary for wartime. "The pavement," a US Federal Works administrator wrote to President Roosevelt, "should be sufficiently heavy to carry mechanized military equipment, and should have shoulders sufficiently wide to permit stopping for repairs or parking without blocking traffic on the paved sections."[24] Ultimately, the PRA recommended that the United States take over the whole construction project and fund a "heavier and better design." As technical specifications were debated in Washington, D.C., the Export-Import Bank put Panama's highway loan on hold pending a decision.[25]

The highway was designated a defense road in 1940 and Roosevelt authorized the allocation of nearly four million dollars to the PRA to construct the Panamanian section, as well.[26] Panama would contribute right-of-way, but no direct funding. The construction of a highway between the seas echoed that of the canal three decades before. Like canal construction, road building required heavy equipment and a large labor force. The PRA, like the canal company, classified labor according to "skill," which they mapped onto national and racial categories. The agency announced that all unskilled labor—"instrument men, chainmen, rodmen, and machete men"—would be Panamanian. By contrast, skilled heavy machinery operators would be imported from the United States, "on account of their experience with the type of equipment used."[27] Finally, Panama's rain, vegetation, and mud made construction a challenge, just as it had during the construction of the canal, rusting equipment designed for drier climates and generally driving up costs.[28]

Construction of the Madden Dam-Cativa section of the highway—to cross twenty-five miles of forests, rivers, and rough topography (figure 10.4)—began in December 1940 when machete men began to "hew away at the jungle" along the route.[29] During the dry months that followed, half

Figure 10.5
Construction of the Transístmica. *Source:* National Archives at College Park, Maryland, RG 30, Box 135.

of the forest along the route was cleared.[30] Then, excavation began. It might not have been the Culebra Cut, but the earthmoving was dramatic: the ground was gashed as deep as one hundred feet, culverts were laid under the roadbed to divert water, and immense soil fills used to plug ravines along the route. As construction progressed, the weather worsened. The dry season ended in April and rainfall built through June and into July, when the rivers swelled and roadwork stopped intermittently.[31] In the months that followed, road crews and machinery stood idle for long periods waiting for the weather to break.[32] Work resumed in January 1942 when the laying of two ten-foot strips of concrete on the Atlantic side commenced.[33] Laborers dumped sacks of cement (which often "magically" disappeared from work sites, pointing to the material's modern allure) into trucks, which fed the paver that laid concrete over the clay roadbed.[34]

A convoy of US army trucks completed the first automobile trip by road between the oceans in Panama on January 21, 1942, after the Atlantic- and Pacific-side construction forces met at the Gatun River.[35] It was clay in sections and paved in others, but an interoceanic road, nonetheless. Paving continued through the next three months with pauses for heavy rains until, finally, a concrete strip connected the Atlantic and Pacific coasts of Panama.

The road was, in a sense, Panama's canal: an interoceanic route that its people could use. "It is possible," the *Panama American* announced on April 17, 1942, "for a car to drive on modern concrete highways from the Pacific to the Atlantic in a little over two hours."[36] On April 15, 1943, almost a year later, the road was formally dedicated as the Boyd-Roosevelt Highway and opened to the public. In his remarks, Panamanian President Ricardo Adolfo de la Guardia used the new concrete highway to link the Panamanian nation, then just four decades old, with the isthmus's long transportation history by suggesting that it was the successor to the Spanish Camino Real road.[37] Drawing a line from mule trains to automobiles, the nationalist discourse around the Transístmica framed its completion as a transformational moment in which a sovereign, forward-looking nation embraced that history on its terms.

The Whole Landscape Turned to Face the Highway

In his description of the first railroad built in colonial Indonesia, historian Rudolf Mrazek observes the magnetic pull of modern transportation infrastructure. He writes, "As soon as rails were laid and the first train appeared, people, the whole landscape, turned around and moved to the train."[38] In much the same way, the landscape turned to face the Transístmica well before its concrete hardened.

In May 1942, nearly a year before the highway was open to the public, President de la Guardia and Panamanian officials conducted a driving inspection with the PRA's chief engineer in Panama, John Humbard. The US government had funded the construction of the road to move military equipment from point to point, but the Panamanian state envisioned the area around the Transístmica as a productive agricultural region with ready access to terminal city ports. And, as the officials drove between the oceans, various state efforts were underway to settle its margins and beyond. The roads department was building dirt penetration roads from the unopened highway to rural communities dependent on water transportation and the government gave settlers plots along the unfinished road.[39]

Figure 10.6
The Transístmica northbound from Panama City, 1950. *Source:* National Archives at
College Park, Maryland, RG 30, Box 136.

A strip of concrete between the oceans. The long-awaited highway
sounds modest in retrospect, but Andrés Díaz, who grew up nearby, told me
that the Transístmica transformed everyday life around the Chagres River.
He and his parents—among the first wave to move to the road—settled
in a town called Buena Vista. Why, I asked, did so many people move to
the new highway town? Correcting me, Andrés said, "Well, it wasn't called
Buena Vista yet. Buena Vista was born around 1947, 1948, or 1949. Before
then it was called Agua Sucia [dirty water] because there is a river there
with that name. There were a few little houses—it wasn't much."
     The highway, he explained, gave the old river town a new name and a
new life. "They [the Canal Zone government] didn't want people farming
in the Zone [after the end of the land lease system]. So what did people
do? Some went to the city. Some went to work in the Zone. Some of us
came to Buena Vista to work in agriculture in the countryside nearby. They
planted bananas along the highway—still a gravel road then—as they had
along the rivers and lakes in past decades. A couple of brothers bought a

truck and became *compradores*, driving up and down the highway buying bananas from farmers and selling them in Colón for export."

Andrés wanted me to understand that the West Indian, Colombian, and Panamanian families that settled in Buena Vista and other highway towns adapted the agricultural practices developed along train lines and waterways to the edges of the road. In this way, the opening of the Transístmica gave rise to an infrastructural ecology that was reminiscent of the landscape that emerged when the canal opened in 1914 and laborers were laid off en masse.

To illustrate the parallels between the opening of the highway and the canal, Andrés transported me back to a time when everything had just opened up, "1914, imagine it! All of these people came to work on the canal, not to farm. When they finished the canal, the people stayed. Options? Bananas! They began to plant bananas. But because of [Canal Zone] restrictions [against farming] people were forced to leave. So they moved to Agua Sucia: 'Yes sir, there's good land there. You can plant bananas and there is a road project coming.'... That is what they did. Sell bananas, drink *aguardiente*, and live happily" (chapter 8).

As the new highway towns boomed, Limón and other lake communities "decayed," many of their inhabitants moved to settle along the road. The primary mode of regional transportation shifted from water to land as the boatmen who had ferried people from the Gatun Lake towns sold their launches and bought small buses called *chivas* to transport people to and from the highway. For the new generation working in Colón or on wartime infrastructure projects around the Panama Canal, the highway made the commute shorter. Meanwhile, for farmers, the edges of the road promised access to *tierra virgin* (virgin land) that was more productive than the banks of the lake and downstream rivers. A land rush began.

Conclusion: Connection as a Precarious Achievement

Why did the entire landscape turn to face the Transístmica? Viewed from a temporal and spatial remove, the human and environmental story that unfolded around the highway has been presented as inevitable. For example, geographer Charles Bennett concluded that the Transístmica "stimulated" the settlement of an isolated area by farmers that, in his words, "invaded" and deforested the region.[40]

This narrative—still prominent in portrayals of regional environmental change—portrays rural history as mechanical process. A road built across arable land has often been understood to "release" social energy or lead to "spontaneous" settlement, but reality is more complex. The world of the highway was developed by the Panamanian state and families' decisions to migrate to the corridor were simultaneously shaped by entrenched structures of landlessness and poverty in the interior and grounded in the complex social and economic realities of their everyday lives.

The construction of the Transístmica was initially opposed by the US government as part of a broader effort to regulate regional movement for public health and defense purposes. For the Panamanian state, by contrast, forests were seen as blocking national development and the roads necessary for its achievement. These infrastructures overlapped at the Transístmica, which doubled as a land route for US military power and a path for national economic development in Panama.

The highway's story suggests that infrastructures channel movement in particular directions, but not in a deterministic way. While white Canal Zone residents traveled its length, rural Panamanians built outward from its edges. Life was quiet and traffic was sparse when people first settled along the Transístmica. Constructed to US military standards, the road was bigger than Panama needed, because the country did not have many automobiles at the time. "When I was a kid," Andrés recalled at the end of our conversation about the Transístmica, "we played baseball in the road. After an hour, one car passed—an hour later, another car. Most of them were Americans—lots of soldiers—but only a few individuals. We played in the road, lied down on it, and, at night, sat and listened to stories."

The road was an altogether different space for white Americans, who traveled between its urban termini rather than over its rural margins. Zonians driving between Panama City and Colón were discouraged from stopping in "unsanitated areas" in Panama after dusk. A Canal Zone circular admonished travelers, "Evening drives are safe only if the automobile is kept at a normal rate of speed and no stop is made." In this way, the highway was not a single object, but a boundary infrastructure that brought together various communities, even as they used and moved over its surface in different ways.

Figure 10.7
Life along the highway, 1950. *Source:* National Archives at College Park, Maryland,
RG 30, Box 136. Panama.

As soon as the Transístmica was opened, it was already becoming something different than what engineers and state officials had envisioned. From the windows of their cars, Zonians would have seen the fringes of a region that was increasingly accessible to and reworked by Panamanians. Together, state institutions, loggers, local transportation providers (*transportistas*), and colonists built out from the highway into the rural areas and created new worlds. Settlement was not, as we will see, an inevitable process stimulated by a "finished" piece of infrastructure. It was a precarious achievement, proceeding through starts, stops, accidents, and reversals. Through hard work, small communities emerged that were partially connected to the highway by dirt roads and capricious waterways.

11   The Conquest of the Jungle

Figure 11.1
Camino de verano built by loggers along the Boquerón River, 1957. *Source:* Smithsonian Institution Archives RU 7006, Box 184, Panama 1955. Used with permission.

The only thing that we need from the government is a road.
—David, Boquerón resident, 2010

By 2008, few residents of Boquerón remembered the details about Macario and the *madereros* (loggers), but most agreed that they cut the first *camino de verano*—a dirt road passable only during the dry season—opening the area around the river to settlement. We know that the road had been carved from the forest by 1957 because the Smithsonian ornithologist Alexander Wetmore photographed it during a collecting trip around Madden Lake that year (figure 11.1).

Loggers were not new to the Chagres River region in the 1950s. In fact, downstream areas along transit routes had been logged for at least a century, dating to the construction of the Panama Railroad in the 1850s.[1] But the flooding of Madden Lake in the 1930s and the opening of the Transístmica in the 1940s made it easier for loggers to move heavy equipment within striking distance of previously inaccessible timber in the river's headwaters. The dirt roads that the loggers cut into the forest from the banks of the lake and the edges of the highway were built as infrastructure for resource extraction, but they also opened the rainy, rugged mountains to migrants from downstream and the interior provinces.

Macario's tractor opened a *corte* (cut) to access the good timber around the Boquerón River during the dry season. It was not much of a road at first—neither wide nor stable enough for the loggers' heavy trucks to use. So they plowed paths down to the river and paid laborers to fell the largest and best trees, which they dragged to the water with their tractors. From there, they floated the logs downstream to barges waiting on Madden Lake. The barges carried the logs to the far end of the lake, near the dam, where they were loaded onto trucks that carried them over the Transístmica to the cities. Logging came to a halt when the deluges of the rainy season began. Over the months that followed, local animal and vehicle traffic would transform the dirt road into impassable mud. When the loggers returned to commence work at the beginning of the next dry season, the road would be destroyed and have to be cut and graded anew.

The loggers were working near the river when the first two families—the Seguras and the Garcías—settled in Boquerón in the dry season of 1958, a year after Wetmore took a photograph of the road. Raúl Segura, his wife Florentina, and their young daughter arrived first. Raúl, who was a middleman in the Transístmica's banana economy during the 1940s, relocated his family from the highway town of Buena Vista, because there was *tierra libre* (unoccupied land) in Boquerón. They built an open-sided structure with a wood frame, thatched roof, and clay floor called a *rancho* near the current townsite, just across the river from the canal's hydrographic station. The Garcías—Chevo and his son Tomás—arrived on foot not long afterward and settled nearby. From the community of Chorrera, some fifteen miles west of Panama City, Chevo and Tomás had, like many other landless campesinos, spent the 1950s moving from one place to the next in search of *tierra libre*. They wanted to cultivate land to which nobody else had a title or possessory right. Like the Seguras, they found that land in Boquerón.

After the Second World War, the Panamanian government channeled landless farmers to forested frontiers as part of its "Conquest of the Jungle" program.[2] State agencies, often supported by loans, programs, and expertise from international development institutions, sought to modernize agricultural production across the rural interior, including the headwaters of the Chagres River, through rural "penetration" roads, agricultural extension, market development, and agricultural credit. In some instances, then, logging roads facilitated regional deforestation by providing pathways for settlers, but they were an informal extension of a multiscale and institutionalized development infrastructure that incentivized extractive relationships between rural people and the land.

The (Always) Unfinished Road

When I began collecting oral histories in Boquerón, I did not ask questions about roads. But the people I interviewed kept bringing them up anyway. After struggling in vain to direct the discussion back to my intended research topic—environmental management—I decided that I should add some new questions to my interview guide to learn more about the road. I asked:

• When did you arrive? What was it like here when you arrived?

• Was there a road here when you arrived? If so, what kind of road?

• When did the construction of the road begin? When was it finished?
• Who built the road?

The questions seemed straightforward, but I could never get the story of the road straight. In their responses, people described the road as always unfinished.

The road had a life of its own, advancing and retreating as changing networks of capitalists, entrepreneurs, and state officials channeled money and labor into the region. Like all infrastructures, Boquerón's road never "arrived" once and for all, because investment in the road and its communities could diminish or dry up completely with a shift in political administrations or commodity prices. Indeed, roads are a matter of concern in Boquerón and many rural communities worldwide precisely because they cannot be taken for granted and so much is at stake.

Scholars have argued that rural people commonly describe poverty in terms of access to infrastructure, particularly roads.[3] Although this corresponds with my ethnographic findings, it also underestimates the potential sophistication of analyses of infrastructural politics and uneven development from the side of the road. People in Boquerón talk about bad roads to characterize more than a general state of poverty. Because road conditions are sociopolitical relationships materialized, the road becomes an object through which community members analyze their changing relationships with—and access to—other groups across space and time.

Why Rural Colonization Was Not "Spontaneous"

"Movement, there was always movement," a friend in Boquerón told me as we sat on the front porch of his hilltop house looking down on the river and the road. "Look," he said, "if people didn't work for the canal, they worked in agriculture, or they did both, moving back and forth between urban and rural places."

The constant movement that he described is political-economic, symbolic, and infrastructural in character. Defined against the historical backdrop of landlessness and poverty in the interior, it is not surprising that wage labor in the cities and the Canal Zone appealed to campesinos. In the past, many poor families in the interior worked on large agricultural estates called *latifundias* controlled by wealthy landowners. They often gained access to small plots to farm for subsistence in exchange for labor on the estates. Given these conditions, farmers migrated for wages or *tierra libre*—their own land. As anthropologist Stephen Gudeman argues, "The fact that the Panamanian peasantry appears to be rootless, to move so often across the landscape—a fact frequently bemoaned in the city—is itself a direct function of this system of land tenure."[4] However, if the political-economic conditions in the interior pushed campesinos to search for land or work elsewhere, they did not determine how, when, and where they moved. In its specifics, migration was shaped by policy, infrastructure, and informal networks.

Campesinos from the Panamanian interior first migrated to the margins of the transit zone in significant numbers in the early 1940s to work on the large construction projects associated with defending the canal during wartime. Construction on two major projects, the Transístmica (chapter 10) and the Third Locks project, began in 1940. The Third Locks project was an effort to expand and make the locks more secure in the face of aerial attack. When the highway was finished and the lock project abandoned, unemployed former laborers from rural backgrounds sought out land in and around the transit zone, where they attempted to recreate agrarian livelihoods, much as they had after the opening of the railroad, bankruptcy of the French canal, and opening of the US canal (chapters 5 and 8).

Neither migration nor settlement patterns were spontaneous. In the 1940s, the Panamanian government created legal incentives for settling frontier areas, including the rural margins of the transit zone. The *patrimonio familiar* (homestead) law of 1941 provided small, titled property concessions (under ten hectares) to landless families who settled on uncultivated private land. The law, the first of its kind, had the greatest impact in the transit zone. A second wartime policy, Decree Law 23 of 1942, extended smallholder colonization by authorizing the occupation of some *latifundias* by landless campesinos. The law was a response to concerns about potential food shortages on the isthmus associated with reduced wartime imports and the disproportionate number of Panamanians working in the construction sector, rather than farming, at that time.[5]

The goal of state initiatives like the "Conquest of the Jungle" was to manage discontent among the rural poor and expand the national economy without disrupting the entrenched political-economic structure of the *latifundia* system that produced inequitable land distribution.[6] For the Panamanian state, the settlement of the rural areas of the transit zone with campesinos served two purposes. First, agricultural production increased food availability in the most populous area of the country. Second, some campesinos settled near the Canal Zone border, provoking confrontations with US authorities. For Panamanian politicians, these David and Goliath clashes provided opportunities to draw attention to the larger issue of US imperialism and national sovereignty.[7] Settlement was focused on the canal's west bank, in the area near Chorrera. The area to the east of the canal was mountainous, rainy and not historically deemed good agricultural land. Also, significantly, it was relatively remote and inaccessible without a road system. However, the opening of the Transístmica provided the infrastructure to better access the headwaters of the Chagres River.

By 1960, Panama's Ministry of Agriculture described the so-called spontaneous colonization of the frontier as an environmental problem. They presented rural ecological ignorance as the cause of deforestation and soil degradation, eliding the historical role of the state in the colonization process. The Ministry suggested state-run agricultural colonies as an appropriate response:

Every year a large number of people, whole families, leave [the interior provinces of] Herrera and Los Santos in search of lands to cultivate, because in these provinces there are none. The families settle in the locations where they are best able, creating with their primitive agricultural practices the same problems that forced them to migrate. In a few years, the children and grandchildren of these families will have to do the same thing, when the lands they occupy today become unusable. It is for this reason that measures should be taken to avoid their continued destruction— which is unnecessary and without benefit—of our forests, wildlife, and water. Well-planned, organized, and administered agricultural colonies with the sentiment of the nation can help to avoid this problem. It will begin with families that voluntarily leave their homes. After, they will be grouped with others so that the successes of the first serve as an example and stimulus for others.[8]

Some colonization was channeled by the formal infrastructure assembled by the state and international organizations to modernize traditional agriculture and extend development into the countryside. Before the 1960s, however, rural development initiatives were limited to minor agrarian reform programs and incentives. The John F. Kennedy administration's Alliance for Progress (1961–1973) provided Panama with forty-one million dollars in assistance for development, six times the amount of external aid received in previous years. The program—which Che Guevara famously described as the "latrinization" of Latin America due to its emphasis on depoliticized infrastructure-led economic development—supported moderate land reforms, small industry, and efforts to improve housing, education, and health across rural areas, but, even with more funds, much of Panama's interior received little attention.[10]

Campesino migration to frontier areas was often called "spontaneous colonization," but that is a misnomer, because their decisions involved complex political and economic calculations. "'Spontaneous colonization,'" sociologist Stanley Heckadon-Moreno writes, "is the name usually given to campesino frontier settlement. Although the term has become widely accepted, it is, unfortunately, misleading. Spontaneity suggests blind impulse or improvisation. From the viewpoint of a poor rural family, however, colonization has nothing to do with an absence of premeditation. Quite the contrary, moving to new lands is considered a highly risky undertaking, requiring careful planning of household activities: acquiring information on the new site's characteristics, marshaling economic resources, and activating a wide network of social ties."[9]

Extending Rural Infrastructure

In the absence of formal state infrastructure, people in Boquerón depended on social networks. The first wave of settlement during the late 1950s and early 1960s worked like this: it began with a "survey," meaning that a rubber tapper, hunter, or logger would travel through the forest. When he made it back to the highway, he spread word about the landscapes that he had seen. Luis Segura, Raúl's son, explained how his father came to Boquerón. "You know how people talk," he said, "'Oye, they cut a road up there and they're starting to work the land.' And one person told another like that." Word spread that the land along the Boquerón River was *montaña virgin* (virgin forest) that was neither settled nor cultivated. The area was not pristine, but the fading marks of centuries of transportation and extraction projects were light on the landscape: the weather-beaten stones of the Camino Real, rubber tapping and hunting trails, remains of an abandoned manganese mine, and some overgrown banana plantations.

Migrating to the frontier was far from easy. First, potential settlers needed to verify the rubber tapper, logger, or hunter's rough description of place. Then, both husband and wife would usually make a reconnaissance trip to the area. For Raúl and Florentina Segura, the fifteen-mile trip from Buena Vista to Boquerón would not have been difficult, but for those considering moves from the provinces of Chiriquí and Veraguas—hundreds of miles away—the journey demanded large amounts of time and money. Once the decision was made, moving the household cost more money. Colonization, then, was too expensive and difficult for the poorest campesinos to afford.[11]

Settlers in Boquerón saw themselves as making a home in the wilderness. People recalled fertile soil, forests full of wild game (deer, iguanas, birds, peccaries, and rodents called *pacas*), and a river full of fish. They said the land was *descansado* (rested) rather than "tired" like the heavily used and degraded soil that they had left behind. Although Boquerón was bountiful, it was also a difficult place to establish a new life. The settlers were unfamiliar with the environment and some people made decisions that reflected their lack of experience. At least one settler family built their house in the river's floodplain, which was swept away by a rainy season flood. Massive trees fell in the wrong direction and killed men cutting them to make fields. Plagues destroyed entire crops. However, the biggest and most enduring problem was a lack of transportation infrastructure.

Boquerón is less than twenty miles away from the Transístmica. Yet the only way to get to the paved highway—and, from there, on to urban centers—was over the logging road or the river. Both routes could be precarious depending upon rainfall. The river was the faster route, but during the dry season some sections were too shallow for boats to pass. The rainy season also made river travel difficult. Higher water allowed boats to travel upstream unimpeded, but extremely heavy rains swelled the river and accelerated its current, forcing travelers to leave their boats on the banks and continue on foot, or wait for the water to recede.

Traveling by land could be even worse. When the loggers left and the rains began, the road turned into a wide swath of mud. During the rainy season, it would take settlers most of a day to walk to the highway without a load, and longer with a pack animal and cargo. Without transportation infrastructure, getting to and from the community was so difficult and time-consuming that many settlers abandoned their farms and

Figure 11.2
Truck load of mahogany off the road in Cerro Azul, Panama, 1949. *Source:* Smithsonian Institution Archives RU 7006, Box 177, Panama 1949. Used with permission.

moved down to the highway or other areas with more accessible roads, electricity, and other utilities. Colonization and the transformation of forested landscapes into a livable place was a precarious achievement, indeed.

In the absence of an all-season gravel or concrete road, community members created an informal transportation infrastructure to facilitate the motion of people and goods. Movement between the highway and the community depended on the *transportistas* with access to motor boats and four-wheel drive trucks.[12] These men—rarely women—knew every pothole in the road and in the river intimately. Their knowledge and labor linked the community to formal transportation networks, making it possible to

Figure 11.3
Packhorse "bogged down on trail" near the Transístmica, 1962. *Source:* Smithsonian Institution Archives RU 7006, Box 187, Panama 1962, Part 1. Used with permission.

get around. While the arrangement was not ideal, it enabled settlers to get crops to market, buy necessities, and access health services.

As the area became more populated, Macario and other loggers continued to work nearby. By the time the first wave of settlers replaced their wooden *ranchos* with cinder block houses, the loggers had cut all of the big trees close to the river and moved deeper into the mountains. The distance between these areas and the river made extracting logs by floating them downstream more difficult, so the loggers improved the road to make it passable in their trucks.

The loggers had no long-term interest in the road per se, or the communities along it. They seasonally maintained and even improved the road (stabilizing it with wooden poles, for example). But, for them, the road was a means to an end. After the good timber was harvested, they would move on. Without the machinery and capital necessary to maintain the road, it would disintegrate and make connection more precarious. Before long, however, Boquerón caught the attention of a new set of actors with even more capital and heavy equipment.

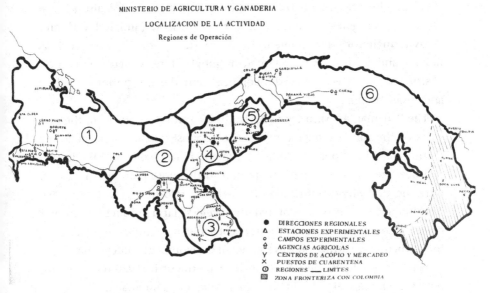

Figure 11.4
By 1970, the upper Chagres River basin was a focus for rural development. *Source*: Government of Panama, *Memoria Comisión de Reforma Agraria* (1970).

The Guardia Nacional and Agrarian Reform

After they took power in 1968, the Guardia Nacional expanded the state's role in rural development as a political and economic project. Rhetorically, their two-pronged approach linked discourses of land reform and national economic development.[13] On the ground, the approach combined infrastructure construction and agricultural programs to modernize rural landscapes and campesino land use practices, drawing the interior into closer relation with a set of domestic and international institutions, banks, development agencies, and commercial interests.

For the Guardia, like other governments across Latin America at the time, tropical forests were spaces to be conquered and civilized. Colonization provided a means of incorporating the forest into the nation-state by exerting territorial control in formerly inaccessible areas.[14] The "Conquest of the Jungle" also worked at the level of discourse. As literary scholar Candace Slater writes, "The jungle is an emphatically non-paradisiacal space. A figurative as well as literal maze ... it is also a place of ruthless struggle for survival. ... Rife with disease ('jungle fever') and decay ('jungle rot'), it is home to beasts and unsavory characters such as hoboes and tramps."[15] Represented as jungle, forests were thought to obstruct national economic development and the integration of urban and rural Panama.

The Panamanian Ministry of Agriculture's annual reports from the first years of the military regime reflect the populist, agrarian reform fervor of the era. The reports proudly document roads built, forests cleared, and new area farmed. Rural penetration roads, more than anything else, were the conduits of settlement and agrarian reform (figure 11.5). They were imbued with promises of making "useless" land productive through colonization and promoted as a means of achieving national integration.[16] Infrastructure construction was a national development strategy that dated to before Panama established its roads department in 1920 (chapter 10), but *asentamientos* suggested a new, more integrated approach.

The Guardia Nacional established the *asentamiento campesino* program in 1969. Modeled on land reform projects developed elsewhere in Latin America, the government established hundreds of small agricultural cooperatives across the country in a political project that involved the expropriation and redistribution of land from *latifundias*.[17] This was a more expansive agrarian reform than its predecessors in Panama. The *asentamientos* provided financial and technical agricultural assistance, developed schools with a vocational emphasis, started new health programs, and constructed low-income housing.[18] The program was aligned with the Guardia's publicly stated—and widely critiqued—priority of improving the condition of the nation's rural poor through land reform.

By many accounts, the cooperatives were met with local skepticism, indifference, or outright hostility. Some campesinos participated in hopes that they would bring better infrastructure to the community. The anthropologist Gloria Rudolf describes a revealing interaction around this issue

Figure 11.5
"How the Agrarian Reform Works." Extending rural infrastructure, 1971. *Source:* Government of Panama, *Memoria Comisión de Reforma Agraria* (1971).

during her 1972 fieldwork in a community she called Loma Bonita in the province of Coclé:

We arrived late. In attendance were eight women and twelve men. Four students (one woman and three men) were sitting at the front of the room. With a tape recorder and a complicated agenda written on the board, students explained that they were studying at the Institute of Inter-American Cooperatives in Panama City. They were in Loma Bonita for about a week to learn about the problems of campesinas/os and to help them. Students divided people into four small groups to discuss community problems. The student leader in my group solicited problems. Silence. In a kind but assertive manner associated with city people, he tried again. This time Geraldo spoke, echoing what I knew to be almost everyone's thoughts. "We need a highway here. Otherwise we can't develop." The student wrote "highway" down on his notepad, but after several others spoke out to support the highway, he explained that a road was actually out of their reach. Anyway, he asked, weren't people interested in a credit cooperative? Everyone shook their heads in agreement as highlanders do under such circumstances. The student wrote "cooperative" at the top of the list and solicited more ideas. When the large group reconvened, to no one's surprise cooperatives were given priority among the issues raised in all groups, while the highway was not mentioned. Several students gave speeches about cooperativism and drew diagrams on the board—tape recording and writing furiously all the while. Another meeting was announced for the next day. As we climbed the path home, everyone was talking about how on earth they would get their agricultural work done during this pivotal season of the year.[19]

State efforts to reorganize the landscapes of rural Panama through infrastructure construction in the late 1960s and 1970s were bound up with a leftist political ideology. However, they also facilitated large-scale capitalist extraction.

Beginning in 1968, the year the Guardia assumed power, Panama's National Cattleman's Association, controlled by the nation's largest livestock owners, lobbied the state to metabolize the forested frontiers for economic development through the establishment of large agricultural loan programs and road construction.[20] The Inter-American Development Bank and National Bank of Panama provided over 90 percent of all loans for livestock production in the Chagres River basin between 1970 and 1979. Meanwhile, local governments, logging companies, the Panamanian military, and Ministry of Public Works built and maintained roads that channeled colonization and amplified deforestation.[21]

Figure 11.6
The Guardia Nacional, Panama's military government, promoted a populist agrarian reform that highlighted the importance of campesinos in social change through slogans like "Revolution is production! Exploit the land, not the man." *Source*: Government of Panama, *Memoria Comisión de Reforma Agraria* (1971).

The Road, the Mine, and the Asentamiento Campesino

By 1970, Panama had 950 miles of paved roads, 705 miles of year-round roads surfaced with gravel and stone, and 2,500 miles of *caminos de verano* usable only during dry weather.[22] At the time, Boquerón's road was still a *camino de verano*, but the community had something that most rural communities hoping for a gravel or paved road did not: manganese. The Guardia Nacional was interested in two mines that dated to the First World War and had been reopened periodically in the decades that followed. The manganese, found in deposits near the river, was used as an alloy to make steel stronger and more workable. In 1915, Jesse Hyatt, a US capitalist, received a mining concession from the Panamanian government and opened a mine under the name Hyatt Panama Manganese Company. Motivated by high manganese prices due to the demand for industrial steel production during wartime, the company built a twenty-six-mile-long narrow-gauge railroad between Boquerón and the town of Portobelo on the Atlantic Ocean. The railroad provided a path into the forests near the river, but its terminus in Portobelo meant that it did not provide ready access to the cities in the same way that the dirt road built by Macario and the loggers to the Transístmica did.

Boquerón's road "arrived" for the second time—as an all-weather gravel *carretera* rather than a dirt *camino*—in the 1970s when the Ministry of Public Works graded the route, laid gravel, and built bridges across ravines so that heavy trucks could come and go, extracting manganese from the mine. In a 1972 annual report, the ministry described road construction efforts in the area as "civic action" projects that were part of a regional modernization effort: "For the first time, the barrier of forgetting [*barrera de olvido*] that has surrounded this province of rich and promising lands has been broken ... the revolutionary government is committed to opening roads where none exist, as is necessary for the development and progress of the country."[23] But, in Boquerón at least, people recognized that the road was built primarily for the manganese, not the community. "Thanks to this manganese, we have a road with a bridge," one female community member told me in an interview, echoing what many others said. That bridge spanned the Diamante ravine, which became the site for a more formal rural infrastructure that was built with the community in mind.

The new gravel road first appeared on maps in the annual reports of state institutions in 1972. That same year, the government founded the Asentamiento el Diamante several miles below the reopened manganese mines and resettled twenty-three landless families from other parts of the country in Boquerón to participate in the project. These state-run agricultural cooperatives were an attempt to control "spontaneous" settlement and create a population for the delivery of social services. The *asentamientos* (planned settlements) were a continuation of decades of efforts by the Ministry of Agriculture to control campesinos and modernize their agriculture, but also reflected leftist politics linked to agrarian reform and focused efforts to develop frontier environments.

A ministry official described the problems facing cooperatives as cultural and infrastructural: "The limited knowledge of the rural man is the largest obstacle to the efficient increase in agricultural cooperatives. Impassable roads during the rainy season is the other negative factor."[24] Decades later, many in Boquerón recalled the human problem somewhat differently, even if they agreed about the bad roads. A longtime resident, whose family was one of the twenty-three that moved to the community in the 1970s to work on the *asentamiento*, summarized its history like this: "Torrijos created the *asentamiento* for people without land. It was a collective work project, meant to unify the people, which I liked. The *asentamiento* provided a lot of benefits, especially medicine, health, and the road. But it was poorly managed, which led to the fall of the organization." The problems that emerged in Boquerón were not unique. As Gloria Rudolf's description of the meeting in "Loma Bonita" above illustrates, cooperative meetings required large amounts of time and took farmers away from actually farming in their *montes*, while not delivering the infrastructure (like roads) that many of the members wanted most. *Asentamientos* often only lasted a few years before dissolving in infighting or because they were used by military officers or other powerful actors to appropriate governmental funds.[25]

Figure 11.7
Asentamiento meetings demanded large amounts of participating farmers' time.
Source: Government of Panama, *Memoria Comisión de Reforma Agraria* (1969).

In the early 1970s, campesinos across the headwaters of the Chagres River inhabited a social world in which, like much of rural Panama, the state actively promoted agriculture and colonization through political, economic, ideological, and infrastructural means. The government administered the Chagres River basin beyond the boundaries of the Canal Zone, where, beginning in the 1960s, they encouraged the transformation of forests into farms through the establishment of a rural development infrastructure of penetration roads, agricultural extension services, schools, and public health posts. As this form of territorial politics changed the landscape, it also forged a moral economy of development in communities. Boquerón's gravel road carried heavy blue government "civic action" trucks bound for the mine and, with them, the promise of a future of economic development and increased participation in national life. When the Asentamiento el Diamante fell apart in 1979, it was less than a decade old. The failed project left behind families and a strong desire for development, especially health services and schools. The road was the physical and symbolic core of an infrastructure that reinforced expectations that the government was committed to maintaining the route and the people living along it, indivisible concerns for rural agriculturalists. The road was supposed to carry the future and then, suddenly, the future changed.

Retracting Rural Development

By 1979, the same year that the *asentamiento* fell apart, the retraction of rural development infrastructure from the recently defined Panama Canal watershed had begun as forest guards and others enrolled in the new regional administration appeared in larger numbers (chapter 3). Not coincidentally, that year also marked the beginning of the transfer of the canal from the United States to Panama. The Guardia Nacional assumed responsibility for regional environmental management around the waterway, and USAID started a multimillion-dollar watershed management program that trained forest guards to regulate land use in communities across the Chagres basin. Downstream, the US government enlisted the Guardia in helping to remove campesino "squatters"—whom the military government had promoted as model political subjects—from the Canal Zone.[26]

The US tropical forester Frank Wadsworth and other early watershed managers perceived rural "culture" and mores to be the root of deforestation around the Panama Canal. However, this narrative elided the role of decades of campesino engagement with a modern physical and financial infrastructure built to transform forests into fields. The upper and lower areas of the watershed were distinguished not only by differences in governmental jurisdiction—United States versus Panama—but also by the territorializing infrastructures that organized populations, landscapes, and subjectivities to serve different purposes.

The problems that watershed managers faced, then, were far more extensive and intransigent than the land use practices of individual campesinos, because they were bound up with long networks of international organizations, state institutions, banks, capitalists, scientific experts, and technologies that stretched tens, hundreds, and even thousands of miles from the Boquerón River and disposed rural people to act in extractive ways toward the land. This infrastructure was organized at the same "level" as the canal. After all, campesinos did not come to Boquerón spontaneously or arrive alone. They came to settle via the road cut by loggers and, later, improved by the government to extract manganese.

Infrastructures are intransigent, giving form to subjectivities and political ecologies that resist rapid change.[27] The watershed could not be integrated on the ground by simply drawing a new boundary on a map or establishing new environmental institutions, because the old development infrastructure retained a momentum that pulled in other directions. In order to extend integrated water management upstream, that had to be changed. To achieve their goals, watershed managers had to both establish new relationships and—equally importantly—disconnect existing networks built to reorganize landscapes for different purposes.

Work stopped at the manganese mine in 1982. When the heavy blue trucks stopped traveling to the mine, road maintenance tapered off. Between 2008 and 2010, community members conducted some maintenance themselves. For example, a work group spent a Saturday rebuilding washed-out bridges with poured concrete and rebar. Because their capacity to complete large maintenance and repair projects is limited by their ability to mobilize capital, technical inputs, and labor, residents seek to make themselves and their road visible to outside politicians in the capitol by enrolling their local *representantes* (representatives).

Conclusion: Infrastructure Expands and Retreats

In oral histories, people in Boquerón described a road that was always unfinished—an emblem of progress that, without constant maintenance, would revert to nature. Roads, like all built infrastructures, require maintenance because they are temporary lines across active environments that erode, rust, and fracture. Maintenance and repair mask this material impermanence, but only temporarily.

Contemporary social theory has predominantly theorized connection and assembly.[28] Boquerón's road suggests that processes of disconnection and disassembly are also critical site of everyday politics and a revealing site for understanding the spatio-temporality of infrastructure. Attention to maintenance and abandonment also suggests that infrastructure is perpetually unfinished. Without constant work—both physical and political—a road can fade into the landscape and the community that depends up on it can become disconnected.

If the history of the Panama Canal watershed is legible to local people in presences like development projects, it is also recognizable through absences on the landscape. During my fieldwork, people recalled neighbors who migrated elsewhere to farm without restriction. They remembered the rise and fall of the *asentamiento* that attracted landless farmers from far away to settle what was then *tierra libre* (free land) on an opening frontier. They pointed to the infrastructure that arrives slowly, if at all. For example, Boquerón is within forty miles of the canal and Panama's largest cities— Panama City and Colón—but electricity first "arrived" in 2010, decades after it had in most of the surrounding region. The gravel road is "improved" but often impassable during the rainy season, leading people to speculate that the infrastructure has been retracted for a reason: to encourage them to move away from the area. In this way, infrastructure renders the logics of distribution embedded in these arrangements visible and raises important questions about the relationship between environmental and social responsibility. Panamanian sovereignty over the territory of the Canal Zone did not necessarily create administrative regimes more responsive to local concerns. Situated within changing global infrastructural relations, environmental policy was, therefore, less a rupture than a continuation and even amplification of US policy.

Part IV   Backwaters

12   Weeds

Figure 12.1
Exterminating water hyacinth plants at Gatun Lake, 1915. *Source:* National Archives at College Park, Maryland, RG 185-G, Box 5, Vol. 9.

What in heaven's name is the reason that the sun never sets on the empire of the dandelion?
—Alfred Crosby, environmental historian, 1986[1]

Water hyacinth (*Eichhornia crassipes*) has been called the world's worst aquatic weed. The invasive floating plant, a native of South America capable of doubling in population and surface area in weeks, is now found in waterways on every continent except Antarctica. In tropical and subtropical areas from Louisiana to Florida, Kenya to Zimbabwe, and India to the Philippines, water hyacinth creates thick mats across the surfaces of lakes, rivers, and canals, choking navigation, hydroelectric production, and other socioeconomic activities that require open waterways. The plant's leaves and attractive flowers are visible above the surface, but it spreads via underwater rhizomes, or stolons, that produce clones and new plant colonies. Hyacinth thrives in slow-flowing fresh water, so invasion often coincides with the alteration of hdyroecology by dams and water management infrastructure, with significant implications for the actors that use those waters.[2]

Water hyacinth control operations are emblematic of the critical, but often overlooked, environmental management and maintenance work that is necessary to move ships across Panama. The Panama Canal's Aquatic Vegetation Control Division manages the large mats of plants that reproduce constantly in the backwaters of the canal—particularly a sluggish stretch of the Chagres River below Madden Dam—and keeps them out of the main shipping channel through the use of engineered (booms on rivers), mechanical (collecting, shredding, and burning), and chemical (pesticide) measures. Engineers estimate that without these measures, the canal would be impassable in three to five years. As the ongoing work of keeping the so-called big ditch free of weeds illustrates, our built infrastructures are more than technological conquests of nature. Instead, they give rise to highly demanding environments that require constant labor and capital inputs to function as designed.[3]

The construction of the Panama Canal has often been characterized as modern man's ultimate conquest of tropical nature. And, without question, canal-related engineering, exvcavation, and sanitation projects profoundly transformed the surrounding region. However, as the water hyacinth problem suggests, the environmental dimensions of transportation are irreducible to the "effects" of the construction project—that is, to conquest. We should also attend to the hybrid nature produced and its lasting implications. The locks and dams built for the canal interacted with webs of social and ecological relationships that were often poorly understood. The managed environments produced for transportation purposes facilitated the geographical expansion of some actors around the river (shipping companies, certain fish, recreationalists, and water hyacinth) while reducing viable habitat for others (big cats, deer, and local farmers).

The water hyacinth problem emerged during canal construction and never went away. When the Chagres River was dammed to provide a year-round water supply for the canal's locks, the river's ecology was transformed from a flowing water (lotic) ecosystem to a still water (lentic) ecosystem (compare figures 5.1 and 6.2). The water hyacinth, a flowering plant kept in the home for ornamental purposes, most likely existed in and around the river before it was dammed. However, the current and flood regime of the river would have swept mats of water hyacinth out to sea before they became too extensive. In the still-to-slow water produced for the lock canal, by contrast, the plant was able to colonize large areas in the absence of human intervention. Let's return to the biological reorganization of the river basin a century ago.

The Canal Zone Biological Survey (1910–1912)

In 1904, the year the United States began canal work, S. P. Langley, secretary of the Smithsonian Institution, contacted President Theodore Roosevelt to inquire about the possibility of conducting scientific work in conjunction with the "digging" of the Panama Canal. His request was broad, characterizing construction as "an opportunity which may never occur again" to gain geological, ethnological, archaeological, biological, and other knowledge of a region being transformed.[4]

Canal engineers and administrators largely saw such research as a distraction from the immediate priorities of construction in terms of time and money. Yet the proposal started to gain traction in 1907 and 1908 after a number of prominent scientific organizations (American Association for the Advancement of Science, American Society of Naturalists, and others) adopted resolutions that emphasized the urgency of completing a biological survey of regional flora and fauna before the flooding of Gatun Lake, which they argued would reorganize or wipe out species. They were most concerned that excavation and the creation of a large lake would facilitate the "mingling" of species—particularly fresh and saltwater organisms—across the continental divide. Therefore, the proposed survey would document species' existence and geographic distribution prior to an epochal environmental shift that threatened to erase geographical differences. It was, in short, a biological salvage project.

Between 1910 and 1912—the peak of construction—a multidisciplinary group of natural scientists from US government institutions and natural history museums participated in the Smithsonian Canal Zone Biological Survey, including botanists Henri Pittier and William R. Maxon, grass specialist Albert S. Hitchcock, entomologists E. A. Schwarz and August Busck, ichthyologists Samuel F. Hildebrand and Seth E. Meek, and mammalogist Edward A. Goldman.

Henri Pittier, the chief botanist, was the first to arrive in Panama. During his first collecting trip in the Canal Zone he observed a "profusion" of flowering plants across a visibly disturbed landscape scarred by railroad tracks, piles of debris, and abandoned town sites. Despite its unnatural appearance, the construction area was rich in plant species and good for collecting. In an enthusiastic letter to the secretary of the Smithsonian, he wrote, "I stopped at the first plant that interested me, and there … within a radius of less than fifty meters, I collected eighty-four species in full blossom. And many more were in sight!"[5]

Pittier observed that the flora of the transit zone, though diverse and uncatalogued, was "secondary and partly adventitious growth." While his first report on Canal Zone plant life traveled by steamship to the Smithsonian in Washington, D.C., the other survey scientists departed New York bound for Panama. Among them was Albert Hitchcock, the biological survey's grass specialist, whose field notes provided some of the best descriptions of the dramatic and often violent environmental transformations taking place around canal works. Upon arrival, he observed, "The whole region is soaked with oil. It forms a scum on pools. Every water course and drainage is black with the destroyed vegetation. The jungle is cut away from along all the lines of drainage and a ditch dug to carry away the water rapidly, these ditches being oiled." Yet, for Hitchcock, the changes in the landscape made his fieldwork easier. "These clearings," he continued, "facilitate collecting for the jungle is impenetrable."[6]

Floating Islands and Worrisome Plants

In August 1911, George Goethals, chief engineer of the canal, summoned Hitchcock to his office. He wanted expert assistance discerning whether aquatic plants seen spreading on the rising Gatun Lake might interfere with navigation through the future canal. Hitchcock traveled to the lake—more of a swamp, actually, since it was just 20 percent full at the time—to investigate:

We went to various points in the launch which drew 3½ feet, and made detailed examinations in a skiff. I was very much surprised to find Para grass (*Panicum barbinode*) growing in 7 feet of water. It was throwing out vigorous stolons on the surface of the water. Below the surface there was a tangled mass of branching runners. I pulled up some that were twenty feet long and an indefinite length in addition. ... At another point I found *Hymenachne auriculata* in dense masses growing in 3 to 4 feet of water and producing vigorous runners. I had not previously observed this species any where on the zone, though it was abundant here. Certain other vegetation was growing in ten feet of water. A common species of *Paspalum*, *P. fasciculatum* was found occasionally here, the point of interest being the runners, which ordinarily have tight sheaths but in this case they ran along the surface of the water, the sheaths being inflated, supporting the runners like bladders. As you will remember this adaptation is conspicuous in the water hyacinth of Florida. ... The grasses in many cases were growing on logs. When the water rose, the logs floated and thus became a centre for a mass of vegetation. ... The masses of vegetation growing in deep water, 6 to 10 feet have produce a substratum extending to the depth of a foot or two below the surface. This is made up of roots, rootstocks and accumulated dirt and debris. ... There was no evidence that the masses were in motion so I suppose they were still anchored by the original stump. This point could not be decided. If they ultimately became detached they would form floating islands. The economic question involved is of course is there danger that these grasses may threaten to interfere with navigation. ... I did not anticipate trouble from this source.[7]

The plants turned out to be more trouble than Hitchcock predicted. Between 1911 and 1913, floating islands up to one hundred feet in diameter spread across the rising lake and its tributaries (figure 12.2), obstructing local boat traffic and raising concerns that they might block navigation through the unfinished canal. The worrisome plants, still unidentified at the time, were concentrated around "floating islands" of debris, which they reportedly bound into a solid masses.[8]

Canal administrators responded to the potential biological threat by pursuing information on the plant's scientific classification, introduction and dispersal in the Zone, as well as treatments employed successfully elsewhere.[9] They located one expert on the isthmus—Dr. Otto Lutz, a professor at the National Institute in Panama—who might be able to identify the plants. Lutz was unable to classify the plant scientifically (he noted that the library of Panama was "badly equipped" for identification).[10] However, he was able to describe its general characteristics and distribution in Central and South America. Lutz reported that the plant was called *oreja de mula* (ear of the mule) in Spanish and water lily or water hyacinth in English. It was kept as an ornamental plant in houses and populated nearly all freshwater bodies on the Atlantic slope from Costa Rica to Colombia, especially stagnant water and shallow lakes. At the Port of Cartagena, he wrote, *oreja de mula* blocked small boats and steamers alike.[11]

Midway through 1913, the US Department of Agriculture confirmed that the aquatic weed in the canal was a close relative of *Eichhornia crassipes*, the floating water hyacinth that was an established problem in slow-moving fresh water from Louisiana to the Nile. By that December, Goethals was receiving numerous reports that water hyacinth was colonizing Gatun Lake's margins and tributary rivers. One report that crossed his desk read, "Water hyacinths in the lake area are increasing very rapidly. From a very few in May these masses have developed both in number and size until they have formed quite an obstacle, and it would seem advisable to begin war upon them as soon as possible."[12]

And so they did. In the United States, federal and state agencies had used two primary methods—chemical and mechanical—to control water hyacinth in Florida and Louisiana. The mechanical method involved dredging, cutting, or otherwise destroying and uprooting the plants. The chemical method, used by the Army Corps of Engineers, involved a spray composed of arsenic, sal-soda, and water that stayed in the water for up to ten days, poisoning the plants. The spray had reportedly been used successfully in Louisiana, but, despite assurances that fish populations would

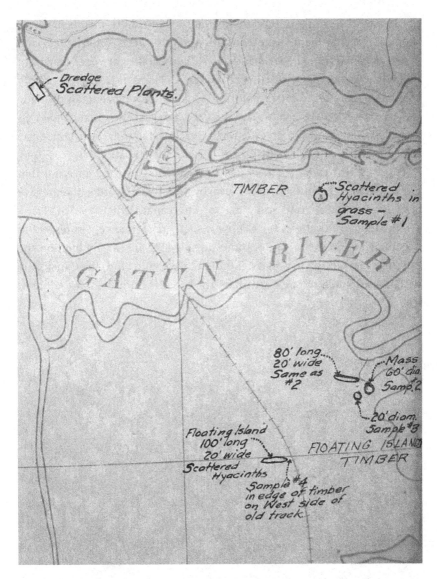

Figure 12.2
Hyacinth plants bound "floating islands" of debris on Gatun Lake, 1913. *Source:*
National Archives at College Park, Maryland, RG 185, Entry 30, File 33.h.4(1).

not be affected, some canal administrators, particularly members of the Sanitation Department, were concerned about effects on "higher" and lower" organisms in the lake, as well as the potability of water drawn for municipal use. Lewis Bates, chief of the canal's Board of Health Laboratory concluded, "From a sanitary point of view the use of arsenic in drinking water is to be condemned."[13] Early in 1914, the year the canal opened, its acting chairman ordered that the water hyacinth be sprayed without delay (figure 12.3). This was the first salvo of a protracted battle.

By 1915, one canal engineer had suggested to the Zone governor that growing water hyacinth be made illegal by "some rule or order or repressive law" in the Canal Zone and Panama. The canal's chief council, Frank Feuille, replied that he did not support such an ordinance. "In respect to Panama," he wrote, "we are without jurisdiction, and in my opinion the interference with the personal rights of people coming into the Canal Zone with plants would not be justified by the benefits from such procedure. It

Figure 12.3
Arsenic-soda spray was used to exterminate water hyacinth, Chagres River, 1925.
Source: National Archives at College Park, Maryland, RG 185-G, Box 5, Vol. 9

is suggested that water hyacinths will probably come down the streams from the Republic into Gatun Lake at all times."[14]

Global Connection and Local Disconnection

Problems with invasive water hyacinth in the Panama Canal have persisted for a century now. The plant has been effectively controlled, but not eliminated, because the canal's locks and dams produce an aquatic environment conducive to its reproduction. Water hyacinth, then, is not simply an environmental problem, but an indicator of how built infrastructures are embedded in ecologies and how that embeddedness redistributes burdens and benefits across human communities.

The canal provided a waterway for interoceanic navigation, but, for decades, it was also the only means of transportation for rural people in the vicinity. In lakeside communities like Limón and the neighboring New Providence, "water cabbage" spread so widely that farmers were unable to travel by boat to their *montes* (farm plots).[15] Moreover, people who lived along nearby tributary rivers, who typically paddled their agricultural products over the lake to the railroad for transport to urban markets, were forced to transport the heavy products miles overland. Residents of lakeside communities asked the canal administration for assistance in exterminating or controlling the hyacinth in canal waters beyond the shipping channel, along the lake shores and rivers traveled by launches and canoes.

The proliferation of water hyacinth underlined the distributional politics of environmental management, particularly tensions around the navigational primacy of the canal and Panamanian use of the environments transformed for shipping. Should the canal administration be responsible for maintaining open water routes for the boats of Panamanian farmers, as well as freighters? In the first decade after the canal opened, administrators conceded that the flooding of Gatun Lake to facilitate interoceanic transportation and global connection had restricted local movement, but debated their government's responsibility to those affected.

Some canal administrators argued that, by changing the flow of the Chagres River, the opening of the canal had produced the conditions that allowed the hyacinth to thrive, thus diminishing local mobility. A memorandum written by Mr. Malsbury, an assistant engineer, in 1919 concluded:

The still water of the lake has made the growth and spread of this weed possible to such an extent that it has become a pest to small craft. Assuming that the river within the particular inlet under consideration was not navigable before the formation of the lake, it might be said that inasmuch as no navigable channel existed prior to inundation no responsibility for the blockage of a channel can rest with the agency responsible for its creation. This argument does not hold, however, for the reason that prior to the formation of the Lake this country was intersected by numerous trails over which the native passed by foot and horse. The lake closed up the trails, leaving only one means of communication to the railroad, ie. cayuca [canoe] transit on the Lake. That the blockade actually exists and that it is due to the water lettuce is beyond question. Mr. F.R. Fitch, who visited this region last week [March 1919], states that it took his party 45 minutes to travel 300 feet. He had a force of ten men, all of whom, himself included, worked to their limit.[16]

Another engineer argued that the commission could not be held responsible for or establish a precedent of compensating those affected by the transformation of the environment for canal purposes. Framing the issue in a manner that still resonates in discussions of transportation and uneven development in Panama today, he concluded that the canal's broader economic benefits on the isthmus outweighed its social costs to affected communities:

I believe that The Panama Canal has no right to assume responsibility or to establish a precedent to remedy possible injuries alleged to be the result of the construction of The Panama Canal. Assuming that the growth of this plant is a result of such construction and that the particular Panamanians in question were injured thereby, it is obvious that such occasional injuries are more than offset by the general good resulting from the Canal construction."[17]

Conclusion: Adventitious Growth

In 1911, the botanist Henri Pittier described the disturbed nature that he saw in the Canal Zone as "adventitious," a designation that turned out to be prescient on two levels. In biology, adventitious refers to a structure that develops in an unusual anatomical position. Rhizomatic hyacinth stolons and banana corms are both adventitious root systems that proliferated in and around the waterway. In general, however, adventitious refers to something that happens by chance, rather than design.[18] At a more abstract register, then, both "social" problems like land conflicts and "environmental" problems like water hyacinth invasion are adventitious because they accreted around, but were not determined by, the Panama Canal's lock design and attendant assembly of infrastructure.

Seeing an iconic canal through a weed illustrates how profoundly our infrastructures are entangled with our political ecologies. It reminds us that engineering marvels have not freed humans from the limitations of the environment, as we once hoped, but bound us more tightly and contentiously to landscapes and waterscapes of our own making. What emerged around the Panama Canal was neither a replica of engineers' blueprints, nor an elevated version of the former riparian world along the lower Chagres River, but something new. By transforming rapid rivers into sluggish lake water, the creation of a lock canal between the oceans produced hydroecologies ripe for plant invasion. As water hyacinth spread around the edge of Gatun Lake and across its feeder rivers, it changed the environment in a manner that became a social and political problem, because it limited regional circulation and development. Thus, the questions of infrastructure, ecology, and responsibility that emerged around the hyacinth spoke—and still speak—to tensions at the heart of the canal project.

13 A Demanding Environment

The road connecting Boquerón to the Transístmica never arrived for good, marking the final conquest of the jungle. Instead, the road was always unfinished, advancing and retreating in relation to the capital, labor, and machinery mobilized by various actors in search of manganese, lumber, and political subjects. For the road to retain the material qualities that defined it as such in the face of the long downpours of the rainy season and heavy traffic, it had to be constantly maintained or the precarious regional connections that it enabled would fall apart.

Rural people are not the only ones in Panama's transit zone who take the impermanence of infrastructure seriously. Canal administrators' interminable struggles with invasive water hyacinth point to a larger truth: the canal's built environment would disintegrate quickly without constant behind-the-scenes maintenance by a multitude of engineers, administrators, and laborers. Its water storage reservoirs would empty as the metal gates that contain them rusted and failed. Sediment would clog undredged channels and the Chagres River would emerge from the canal and follow its old channel north to the Atlantic Ocean. As the Pacific section of the Culebra Cut—where most excavation took place a century ago—dried up, North and South America would be reunited.[1]

Infrastructures, then, are not permanent but processual. They require human communities to maintain them, even as they shape those (and other) communities. Without maintenance, infrastructures crack, rust, and crumble and the political projects, promises, and aspirations that they carried dissipate as formerly connected places are disconnected. Maintenance is necessary because infrastructures—even global infrastructures—are both networked and embedded in the environments they cross and transform. Dams and reservoirs, for example, are generic features of lock canals designed to insulate them from fluctuations in weather and climate. They reorganize rivers, appearing to make them predictable service providers—just another part of the system. Yet the nonhuman environment resists complete incorporation into infrastructure. If one of the ecological byproducts of the Panama Canal is an invasive weed that threatens to run amok and, therefore, requires new layers of control and maintenance, then its social byproducts include human communities whose histories are entangled with that of the waterway. Today, moving ships means managing political ecologies that, like weeds, have grown up around the canal.

The opening of the Panama Canal was once held up as modern man's ultimate triumph over nature. It became a key node along a global circulatory system that changed human relationships with the environment, but not only in the ways anticipated. Rather than emancipating humanity from natural constraints, modern technologies of production and communication bound the global population more tightly and precariously to what seemed to be a shrinking planet. In 1972, the anthropologist Clifford Geertz, who had spent years studying human-environment relationships in villages in the so-called third world wrote,

It used to be thought that, although environment might shape human life at primitive levels, where men were, it was said, more dependent on nature, culture-evolutionary advance, especially technical advances, consisted of a progressive freeing of man from such conditioning. But the ecological crisis has divested us all of that illusion; indeed, it may be that advanced technology ties us in even more closely with the habitat we both make and inhabit, that having more impact upon it we in turn cause it to have more impact upon us.[2]

His observation that advanced technology binds us to a world of our own making is still appropriate for the infrastructure-environment problems we face today.

Infrastructures give rise to *demanding environments*.[3] Depending on who, where, and when you are, they can produce different experiences of control over or separation from the nonhuman world, while increasing vulnerability to its variability.[4] Engineered technologies are embedded with historical assumptions about economy, society, and environment, which they carry into worlds in flux. Consider the changing answers to these questions: How much rain will fall across the Chagres River basin in a given year? How many ships will transit the canal every day? How large will those ships be? How many laborers will be needed? When? What water demands—besides shipping—will be placed on the discharge regime of the river?

Whether organized around transportation, agriculture, forestry, or finance, the modern single-purpose environment is highly demanding.[5] For example, the failures of the green revolution demonstrated that once we alter a landscape to serve commodity agriculture production through modified seeds, irrigation systems, fossil fuel–based fertilizers, tractors and combines, and transportation systems to get products to market, we have constructed a landscape that, like the Panama Canal environment, would fall apart in short order if the technologies assembled to make it perform were removed or not maintained. Once demanding environments have been established, then, there is great impetus to continue and even expand maintenance investment against a tendency of things to fall apart.

Landscape architects and urban design scholars have critiqued the centralized, technocratic, monofunctional approach to infrastructure dominated by civil engineers. The infrastructures of the twentieth century were siloed in terms of design and departmental administration, often making them inflexible as social and economic conditions changed.[6] One way forward, then, is to replace aging engineered hardware with polyfunctional landscape infrastructures that provide multiple environmental services. While this shift is promising, we must keep in mind that remaking landscapes as infrastructure, like building dams and roads, inevitably produces winners and losers.

I introduced the book by challenging the popular notion that the Panama Canal is a "big ditch" between the Atlantic and Pacific completed in 1914. The canal casts a long shadow across regional communities and landscapes that have become part of the ecology of global transportation, but which are often invisible to its far-flung users. Yet the canal cannot operate independent of its social and environmental context. It needs water and workers, pumps and policies. The relationships between the canal and surrounding communities are not simply relations of dominance, but interdependence. Human emotions, memories, and feelings may seem peripheral to the of moving ships, but maintaining an infrastructure that is part of the landscape depends on addressing different experiences with infrastructures, both those materially present and those that have come and gone, leaving behind only promises and expectations.

Notes

Preface

1. Clifford Geertz, "The Wet and the Dry: Traditional Irrigation in Bali and Morocco," *Human Ecology* 1, no. 1 (1972): 38.

Chapter 1

1. This estimate is based on 2010 figures provided to the news agency *EFE* by Panama's national authority of public services that place per capita consumption of potable water in Panama at 106 gallons/day, the highest in Latin America. Meanwhile, as the article notes, 16 percent of the population has no access to potable water. See "Panamá es el Mayor Consumidor de Agua per Cápita en América Latina," *EFE*, March 22, 2010.

2. Frank Wadsworth, "Deforestation: Death to the Panama Canal," in *US Strategy Conference on Tropical Deforestation* (Washington, DC: US Department of State and US Agency for International Development, 1978), 22–25.

3. Stanley Heckadon-Moreno, "Impact of Development on the Panama Canal Environment," *Journal of Interamerican Studies and World Affairs* 35, no. 3 (1993): 138.

4. This approach is epitomized by David McCullough's magisterial *The Path Between the Seas: The Creation of the Panama Canal, 1870–1914* (New York: Simon and Schuster, [1977] 2001). See also Gerstle Mack, *The Land Divided: A History of the Panama Canal and Other Isthmian Canal Projects* (New York: Knopf, 1944); and Matthew Parker's more critical *Panama Fever: The Battle to Build the Panama Canal* (London: Hutchinson, 2007). For postconstruction works by what are arguably the two best-known canal administrators, see George Goethals, *Government of the Canal Zone* (Princeton, NJ: Princeton University Press, 1915); and William Gorgas, *Sanitation in Panama* (New York: D. Appleton and Company, 1915).

5. In English, see Michael L. Conniff, *Black Labor on a White Canal: Panama, 1904–1981* (Pittsburgh, PA: University of Pittsburgh Press, 1985); Julie Greene, *The Canal Builders* (New York: Penguin, 2009); John Lindsay-Poland, *Emperors in the Jungle: The Hidden History of the US in Panama* (Durham, NC: Duke University Press, 2003); and Paul S. Sutter, "Nature's Agents or Agents of Empire? Entomological Workers and Environmental Change During Construction," *Isis* 98, no. 4 (2007): 724–754. Historian Walter LaFeber describes the US presence in Panama as informal colonialism. He writes, "The colonial ties were informal; that is, they were not systematically designed as a colonial system to be operated (as it was in London) by a Colonial Office. Panama ostensibly retained independence and sovereignty. Yet the North American control of Panama [was characterized by] relationships developing less formally, less systematically and handled by the State Department rather than a Colonial Office." Walter LaFeber, *The Panama Canal: The Crisis in Historical Perspective* (New York: Oxford University Press, 1978), 67–68. In Spanish see Alfredo Castillero Calvo, "Transistmo y Dependencia: El Caso del Istmo de Panamá." *Loteria* 211 (1973): 25–56; Guillermo Castro Herrera, *El Agua Entre los Mares* (Panama City: Ciudad de Saber, 2007); Omar Jaén Suárez, *Análisis Regional y Canal de Panamá: Ensayos Geográficos* (Panama City: Editorial Universitaria, 1981); Jorge Mastellari Navarro, *Zona Del Canal: Analogía de Una Colonia* (Panama City: n.p., 2003).

6. Chandra Mukerji, *Impossible Engineering: Technology and Territoriality on the Canal Du Midi* (Princeton, NJ: Princeton University Press, 2009), 226.

7. Panama Canal Authority, *Annual Report*, 2012.

8. Ibid.

9. Andrew Barry, "Technological Zones," *European Journal of Social Theory* 9, no. 2 (2006): 243–244.

10. The seminal work of world-system theory is Immanuel Wallerstein, *The Modern World-System I: Capitalist Agriculture and the Origins of the European World-Economy in the Sixteenth Century* (New York: Academic Press, 1974). For a useful summary, see Peter Taylor "World-System Theory," in *The Dictionary of Human Geography*, ed. R.J. Johnston, Derek Gregory, Geraldine Pratt, and Michael Watts (Malden, MA: Blackwell Publishing, 2000), 901–903. Key works of dependency theory include Andre Gunder Frank, *Capitalism and Underdevelopment in Latin America* (New York: Monthly Review Press, 1969); and Raul Prebisch, *International Economics and Development* (New York: Academic Press, 1972).

11. See Arjun Appadurai, "Disjuncture and Difference in the Global Cultural Economy," in *Modernity at Large: Cultural Dimensions of Globalization* (Minneapolis: University of Minnesota Press, [1990] 1996), 27–47; Manuel Castells, *The Rise of the Network Society* (Hoboken, NJ: Wiley, 1996); and Ulf Hannerz, "Notes on the Global Ecumene," *Public Culture* 1, no. 2 (1989): 66–75.

12. James Ferguson and Akhil Gupta, "Spatializing States: Toward an Ethnography of Neoliberal Governmentality," *American Ethnologist* 29, no. 4 (2002): 982.

13. This critique of globalization and global theory is articulated eloquently in James Ferguson, *Global Shadows: Africa in the Neoliberal World Order* (Durham, NC: Duke University Press, 1999), 42–48; Stuart Alexander Rockefeller, "Flow," *American Anthropologist* 52, no. 4 (2011): 557–578; and Christine J. Walley, *Rough Waters: Nature and Development in an East African Marine Park* (Princeton, NJ: Princeton University Press, 2004), 6–12, 264.

14. These frameworks are developed in Andrew Barry, "Technological Zones"; Stephen Graham and Simon Marvin, *Splintering Urbanism: Networked Infrastructure, Technological Mobilities and the Urban Condition* (London: Routledge, 2001); Bruno Latour, *Reassembling the Social: An Introduction to Actor-Network-Theory* (Oxford: Oxford University Press, 2005); Stephen J. Collier and Aihwa Ong, "Global Assemblages, Anthropological Problems," in *Global Assemblages: Technology, Politics, and Ethics as Anthropological Problems*, ed. Aihwa Ong and Stephen J. Collier (Malden, MA: Blackwell, 2005), 3–21; and Anna Tsing, *Friction: An Ethnography of Global Connection* (Princeton, NJ: Princeton University Press, 2005). On space, science, and technology, see also John Law, "Objects and Spaces," *Theory, Culture & Society* 19, no. 5/6 (2002): 91–105; John Law and Annemarie Mol, "Situating Technoscience: An Inquiry into Spatialities," *Environment and Planning D: Society and Space* 19, no. 5 (2001): 609–621.

15. The history of the word comes from the Oxford English Dictionary, "Infrastructure," Oxford: Oxford University Press, (1991), 950; The quotation from the 1927 Oxford English Dictionary comes from William H. Batt, "Infrastructure: Etymology and Import," *Journal of Professional Issues in Engineering* 110, no. 1 (1984): 2.

16. Geoffrey C. Bowker, *Science on the Run: Information Management and Industrial Geophysics at Schlumberger, 1920–1940* (Cambridge, MA: MIT Press, 1994), 10.

17. "Infrastructure studies" is largely undefined as a field. The scholars that I draw on here—Geoffrey Bowker, Paul Edwards, and Susan Leigh Star—are associated with science and technology studies and have worked in departments of information science, informatics, and computer science. They have focused empirically on information infrastructure, while retaining a broader interest in the relationships between infrastructure and society. See Geoffrey C. Bowker and Susan Leigh Star, *Sorting Things Out: Classification and Its Consequences* (Cambridge, MA: MIT Press, 1999); Paul Edwards, "Infrastructure and Modernity: Force, Time, and Social Organization in the History of Sociotechnical Systems," in *Modernity and Technology*, ed. Thomas J. Misa, Philip Brey, and Andrew Feenberg (Cambridge, MA: MIT Press, 2003), 185–226; Paul Edwards, Steven J. Jackson, Geoffrey C. Bowker, and Cory P. Knobel, *Understanding Infrastructure: Dynamics, Tensions, and Design/Report of a Workshop on History and Theory of Infrastructure: Lessons for New Scientific Cyberinfrastructures* (Ann

Arbor, University of Michigan, 2007); Susan Leigh Star, "The Ethnography of Infrastructure," *American Behavioral Scientist* 43, no. 3 (1999): 377–391; Susan Leigh Star and Karen Ruhleder, "Steps Toward an Ecology of Infrastructure: Design and Access for Large Information Spaces," *Information Systems Research* 7, no. 1 (1996): 111–134.

18. On the gateway concept, see Edwards et al., *Understanding Infrastructure*; and Tineke Egyedi, "Infrastructure Flexibility Created by Standardized Gateways: The Cases of XML and the ISO Container," *Knowledge, Technology & Policy* 14, no. 3 (2001): 41–54. As Lawrence Busch shows, nineteenth-century railroad engineers selected different gauges—the distance between the rails—in system designs around the world. Railroads began to replace inland water navigation, incentivizing intersystem coordination (there were nine different gauges in use in the United States alone before the 1880s). The process of establishing standard gauges without gateways was costly and time-consuming. Electrical, communications, and other transportation systems followed a similar pattern, shifting from an emphasis on construction to coordination. See Lawrence Busch, *Standards: Recipes for Reality* (Cambridge, MA: MIT Press, 2011), 64–65.

19. Barry, "Technological Zones."

20. Marc Levinson, *The Box: How the Shipping Container Made the World Smaller and the World Economy Bigger* (Princeton, NJ: Princeton University Press, 2006). On labor and shipping containers, see Busch, *Standards*, 166. The concept of standard containers as gateways is developed in Egyedi, "Infrastructure Flexibility Created by Standardized Gateways."

21. Levinson, *The Box*, 233.

22. Brian Larkin, "The Politics and Poetics of Infrastructure," *Annual Review of Anthropology* 42, no. 1 (2013): 327–343. Larkin observes that infrastructures operate on multiple levels concurrently. They serve technical functions (moving water, electricity, or ships) and as representational forms through which states, corporations, and other entities produce meaning at a level of fantasy and desire.

23. Tsing, *Friction*, 55–65.

24. See Batt, "Infrastructure," on the intertwined history of infrastructure, international economic development, and global military coordination after the Second World War. The classic work of modernization theory that popularized the idea of "stages of growth" and coined the term "take off" is Walter W. Rostow, *The Stages of Economic Growth* (Cambridge: Cambridge University Press, 1960). For an overview of development theory, see Richard Peet and Elaine Hartwick, *Theories of Development: Contentions, Arguments, Alternatives* (New York: Guilford Press, 2009).

25. Busch, *Standards*, 168–169.

26. Paul Edwards has called this cross-scale approach to infrastructure "mutual orientation." See Edwards, "Infrastructure and Modernity," 186.

27. Historian Sara B. Pritchard points out that social and historical studies of technology tend to "black-box" the nonhuman environment, treating it as an unproblematic and static backdrop for technological development; Sara B. Pritchard, *Confluence: The Nature of Technology and the Remaking of the Rhone* (Cambridge, MA: Harvard University Press, 2011), 12. However, a growing body of work examines the coproduction of environment and technology, ranging from envirotech approaches in history; actor network theory; hybrid and posthumanist anthropologies, geographies, and philosophies; and political ecology. See, for example, Jane Bennett, *Vibrant Matter: A Political Ecology of Things* (Durham, NC: Duke University Press, 2010); Timothy Mitchell, "Can the Mosquito Speak?," in *Rule of Experts: Egypt, Techno-Politics, Modernity* (Berkeley: University of California Press), 19–53. Peter Redfield, *Space in the Tropics: From Convicts to Rockets in French Guiana* (Berkeley: University of California Press, 2000); Martin Reuss and Stephen H. Cutcliffe, *The Illusory Boundary: Environment and Technology in History* (Charlottesville: University of Virginia Press, 2010); Sarah Whatmore, *Hybrid Geographies: Natures, Cultures, Spaces* (London: SAGE, 2002); and Richard White, *The Organic Machine: The Remaking of the Columbia River* (New York: Hill and Wang, 1995).

28. See Michael Fischer, "Technoscientific Infrastructures and Emergent Forms of Life: A Commentary," *American Anthropologist* 107, no. 1 (2005): 55–61.

29. William Cronon, *Nature's Metropolis: Chicago and the Great West* (New York: W.W. Norton and Company, 1991). Other influential work rethinking cities through infrastructure, territory, and ecology includes: Mike Davis, *Ecology of Fear: Los Angeles and the Imagination of Disaster* (New York: Vintage, 1999); Matthew Gandy, *Concrete and Clay: Reworking Nature in New York City* (Cambridge, MA: MIT Press, 2002); and Erik Swyngedouw, *Social Power and the Urbanization of Water: Flows of Power* (New York: Oxford University Press, 2004). On understanding the stories of places by following "paths out of town," see William Cronon, "Kennecott Journey: The Paths Out of Town," in *Under an Open Sky: Rethinking America's Western Past*, ed. William Cronon, George Miles, and Jay Gitlin (New York: W. W. Norton and Company, 1992), 28–51.

30. For foundational work on territoriality in geography, see Robert David Sack, *Human Territoriality: Its Theory and History* (Cambridge: Cambridge University Press, 1986); and Peter Vandergeest and Nancy Lee Peluso, "Territorialization and State Power in Thailand," *Theory and Society* 24, no. 3 (1995): 385–426; On the materiality of the state, expertise, and territory, see Tony Bennett and Patrick Joyce, *Cultural Studies, History and the Material Turn* (New York: Routledge, 2010); Patrick Carroll, *Science, Culture, and Modern State Formation* (Berkeley: University of California Press, 2006); Mukerji, *Impossible Engineering*; Penelope Harvey, "The Materiality of State Effects: An Ethnography of a Road in the Peruvian Andes," in *State Formation:*

Anthropological Perspectives, ed. Christian Krohn-Hansen, Knut G. Nustad (London: Pluto Press, 2005), 216–247; Timothy Mitchell, *Carbon Democracy: Political Power in the Age of Oil* (London: Verso, 2013); and James C. Scott, *Seeing Like a State: How Certain Schemes to Improve the Human Condition Have Failed* (New Haven, CT: Yale University Press, 1998).

31. Mukerji, *Impossible Engineering*, 5.

32. Ibid., 207.

33. On entangled landscapes, or sites where "multiple spatialities, temporalities, and power relations combine," see Donald S. Moore, *Suffering for Territory: Race, Place, and Power in Zimbabwe* (Durham, NC: Duke University Press, 2005), 4. As anthropologist Peter Redfield points out, studies of science and technology have been particularly useful in helping us locate and describe the centers of modern practice, but have paid less attention to their edges; Peter Redfield, "Beneath a Modern Sky: Space Technology and Its Place on the Ground," *Science, Technology, & Human Values* 21, no. 3 (1996): 251–274.

34. On empire and territory, see Mark Gillem, *America Town: Building the Outposts of Empire* (Minneapolis: University of Minnesota Press, 2007), 3–4. For discussion of cultural categories and the intimacies of colonialism and imperialism, see Gilbert M. Joseph, Catherine C. LeGrand, and Ricardo D. Salvatore, *Close Encounters of Empire: Writing the Cultural History of US-Latin American Relations* (Durham, NC: Duke University Press, 1998); and Ann Laura Stoler, "Tense and Tender Ties: The Politics of Comparison in North American History and (Post) Colonial Studies," *The Journal of American History* 88, no. 3 (2001): 829–865.

35. Hay–Bunau-Varilla Treaty (Panama Canal Treaty) of 1903, Article II.

36. I thank Kurt Dillon for this observation about the history of enclaves.

37. The water clause appears in the Hay–Bunau-Varilla Treaty (Panama Canal Treaty) of 1903, Article IV. On the past and present of water politics on the Panamanian isthmus, see Guillermo Castro Herrera, "Panamá: Agua y Desarrollo en Vísperas del Segundo Siglo," *Revista Tareas* 114 (May–August 2003): 21–52.

38. As urban Panama develops and the economy grows at a rapid pace, the question of economic disintegration remains important. According to Index Mundi's *Panama Economy Profile 2013*, the service sector contributes 79 percent to national economic activity, dwarfing manufacturing at 17 percent and agriculture at 4 percent, but yet 17 percent of the population works in agriculture.

39. On chokepoints, shipping, and the Panama Canal, see Jean-Paul Rodrigue, "Straits, Passages and Chokepoints: A Maritime Geostrategy of Petroleum Distribution," *Cahiers de Géographie du Québec* 48 (December 2004): 357–374. I analyze the historical construction of and infrastructural inertia behind Panama's "natural advantage" as a transportation provider in chapter 5 of this book.

40. Levinson, *The Box*, 234–235.

41. Prior to the referendum, Panamanian critics questioned the "real cost" of the expansion and historically inequitable distribution of transportation wealth.

42. Andrea Hricko, "Progress and Pollution: Port Cities Prepare for the Panama Canal Expansion," *Environmental Health Perspectives* 120, no. 12 (2012): A471.

43. Mim Whitefield, "Maritime execs debate Panama Canal expansion at Miami conference," *Miami Herald* (October 2, 2013).

44. Bowker, *Science on the Run*, 104; Paul Edwards, *A Vast Machine: Computer Models, Climate Data, and the Politics of Global Warming* (Cambridge, MA: MIT Press, 2010), 20.

Chapter 2

1. Throughout this book, I have changed the names of the people I interviewed to protect their identities. Following convention, the names of historical figures and contemporary public figures, like politicians, have remained unchanged. Place names are also unaltered because it would be difficult to hide their identity and specific geographic locations are important to my arguments.

2. "Flood Warning System on Rivers Tributary to Proposed Madden Lake," *Canal Record*, September 15, 1933.

3. Stanley Heckadon-Moreno, "Panama's Expanding Cattle Front: The Santeno Campesinos and the Colonization of the Forests" (Dissertation, Department of Sociology, University of Essex, 1984), 196.

4. On the Panamanian *roza* agricultural calendar, see Heckadon-Moreno, "Panama's Expanding Cattle Front," 198–204; Francisco Herrera, *Análisis de Los Datos Obtenidos Por RENARE En El Área de Estudio En Octubre de 1982*, (Panama City: RENARE, 1984), 27–29; Gloria Rudolf, *Panama's Poor: Victims, Agents, and Historymakers* (Gainesville: University Press of Florida, 1999), 41; and Stephen Gudeman, *The Demise of a Rural Economy: From Subsistence to Capitalism in a Latin American Village* (London: Routledge, [1978] 1988), 66–72.

5. This description is based on my semistructured interviews and informed by the analysis of the concepts of *monte, rastrojo,* and agricultural production in Panama as presented in Gudeman, *The Demise of a Rural Economy*, 70–71; and Heckadon-Moreno, 1984, "Panama's Expanding Cattle Front," 196, 283.

6. Stanley Heckadon-Moreno, "La Ganadería Extensiva y La Deforestación: Los Costos de Una Alternativa de Desarrollo," in *Agonía de La Naturaleza*, ed. Stanley Heckadon-Moreno and Jaime Espinos González (Panama City: Instituto de Investigación Agropecuaria de Panamá, 1985); Stanley Heckadon-Moreno, *Cuando se Acaban*

Los Montes: Los Campesinos Santeños y la Colonización de Tonosi (Panama City: Editorial Universitaria Carlos Manuel Gasteazoro, 2006).

7. Gudeman, *The Demise of a Rural Economy*, 61.

Chapter 3

1. Author interview with Stanley Heckadon-Moreno, October 8, 2009.

2. In this chapter, I use the terms "watershed" and "drainage basin" interchangeably. "Watershed" entered English at the beginning of the nineteenth century from the German *wasserscheide*, or water-parting. The English usage of the term, like the German, originally referred to the boundary line dividing drainage basins. By the late-nineteenth century, however, watershed increasingly referred to "the whole gathering ground of a river system," which is how it is used in this chapter. The watershed is perhaps the key unit in contemporary environmental management and planning, but in-depth studies of the scientific and political history of the concept are surprisingly rare. See "Watershed" in *Oxford English Dictionary*, ed. J. A. Simpson and E. S. C. Weiner (Oxford: Oxford University Press, 1991); T. C. Smith, "The Drainage Basin as an Historical Unit for Human Activity," in *Introduction to Geographical Hydrology*, ed. R. J. Chorley (London: Methuen, 1971), 20–29; Ludwik A. Teclaff, *The River Basin in History and Law* (The Hague: Martinus Nijhoff, 1967); and Francois Molle, "River-Basin Planning and Management: The Social Life of a Concept," *Geoforum* 40, no. 3 (2009): 484–494.

3. This distinction was legally established in the Republic of Panama's Forest Law 13 of 1987. The law prohibited cutting of all primary and secondary forest (i.e., *rastrojo*) more than five years old. The law was implemented and enforced by INRENARE, the national environmental management agency that was the predecessor of ANAM. For a discussion of *rastrojo*, see chapter 2 in this book.

4. The Republic of Panama first declared the Chagres National Park in 1984 through the *Decreto Ejecutivo 73 de 2 de Octubre* (Executive Decree 73 of October 2). The park was formally established in 1985 with the publication of that executive decree in the *Gaceta Oficial* 20.238 (Official Gazette 20.238).

5. Frank Wadsworth, "Deforestation: Death to the Panama Canal," in *US Strategy Conference on Tropical Deforestation* (Washington, DC: US Department of State and US Agency for International Development, 1978), 23.

6. Ibid.

7. On drought and water supply during this period, see W. F. Bullock, "The Panama Canal, Problems that It Presents, Droughts and Floods, the Time It Saves," *The Mercury*, March 24, 1932; and "Two-Year Drought Hits Panama Canal," *New York Times*, January 25, 1931. On the construction of Madden Dam and water supply

concerns, see R. Z. Kirkpatrick, "Madden Dam Will Insure Water Supply for Gatun Lake Development," *Iowa Engineer* (March 1934): 84–85;

8. For a summary of the dynamic relationships among ship size, traffic volume, and the limitations of dry season water availability within the Panama Canal, see "Gatun Lake Bed May Be Deepened," *New York Times*, November 24, 1967.

9. "Draft on Panama Canal: Low Water Level Leads to Limitation on Big Tankers and Ore Carriers," *New York Times*, June 28, 1957; Olive Brooks, "Long Panama Canal Dry Season Finds Aides Glad for Wet Guess," *New York Times*, July 27, 1957.

10. "Dry Season Curbs Panama Shipping," *New York Times*, April 19, 1961; "A New Dam Project Is Studied to Increase Capacity in Panama," *New York Times*, October 7, 1962.

11. "Canal Transits at Peak 2nd Straight Month," *Panama Canal Spillway*, March 13, 1964; "Limitations on Draft Effective," *Panama Canal Spillway*, March 10, 1964; "When Will Late Rainy Season Start?," *Panama Canal Spillway*, April 2, 1965; "March Transits Set New Record," *Panama Canal Spillway*, April 9, 1965; "Canal Traffic at New Highs," *Panama Canal Spillway*, April 23, 1965; "Low Gatun Lake Restricts Transits of Panama Canal," *New York Times*, April 25, 1965; "Panama Canal Cuts Ships' Draft Limit," *New York Times*, July 25, 1965.

12. Wadsworth, "Deforestation: Death to the Panama Canal," 23.

13. Donald J. Pisani, "A Conservation Myth: The Troubled Childhood of the Multiple-Use Idea," *Agricultural History* 76, no. 2 (2000): 155.

14. The sponge effect is a controversial formulation at the time and remains so today. See Leendert A. Bruijnzeel, "Hydrological Functions of Tropical Forests: Not Seeing the Soil for the Trees?," *Agriculture, Ecosystems and Environment* 104, no. 1 (2004): 186; James S. G. McCulloch and Mark Robinson, "History of Forest Hydrology," *Journal of Hydrology* 150, no. 2 (1993): 189–216; Vasant K. Saberwal, "Science and the Desiccationist Discourse of the 20th Century," *Environment and History* 4, no. 3 (1998): 309–343. McCulloch and Robinson identify three prominent historical myths about forest-water relationships: (1) forests "make" rain, (2) forests reduce floods and erosion, and (3) forests augment low/dry season flows. These "myths" remain, to varying degrees, contentious or unresolved issues in contemporary forest hydrology. First, there is no experimental evidence of increased rainfall following the conversion of bare or cultivated land into forest. See Lawrence S. Hamilton and Peter N. King, *Tropical Forested Watersheds: Hydrologic and Soils Response to Major Uses or Conversions* (Boulder, CO: Westview Press, 1983). Second, the effects of forest cover on erosion remain inconclusive. While natural forests seem to limit erosion, the effect appears to be site- and species-specific. See Ian Calder, "Forests and Water—What We Know and What We Need to Know," *Science for Nature Symposium* (Washington, DC, 2006). Finally, there is no scientific consensus on the relationship

between forests and low/dry season flows. See McCulloch and Robinson, "History of Forest Hydrology." Forests improve water infiltration and replenish groundwater reserves, which should translate to greater water availability in the dry season. However, a review of ninety-four paired watershed experiments found no evidence that the reduction of forest cover led to reductions in water yield, nor any in which increases in cover led to increases in yield. See J. M. Bosch and J. D. Hewlett, "A Review of Catchment Experiments to Determine the Effect of Vegetation Changes on Water Yield and Evapotranspiration," *Journal of Hydrology* 55, no. 1 (1982): 16. Which of these effects dominates seems to be contextual, depending on a combination of factors including rainfall regime, soil type, and land use. But they are still potent regardless when invoked in political discourse. See David Kaimowitz, "Useful Myths and Intractable Truths: The Politics of the Links between Forests and Water in Central America," in *Forests, Water and People in the Humid Tropics: Past, Present and Future Hydrological Research for Integrated Land and Water Management*, ed. Michael Bonnell and Leendert A. Bruijnzeel (Cambridge: Cambridge University Press, 2004), 86–98. In the twentieth century, foresters acquiesced to the political use of some questionable claims about hydrologic relationships because the rhetoric was used to protect forests. Lawrence Hamilton and Peter King have cautioned that, while well intentioned, such claims may produce a backlash against forestry and conservation if watershed forests are protected but flooding, droughts, and waterway siltation continue. See Hamilton and King, "Tropical Forest Watersheds," 131.

15. Scholars suggest that Pliny the Elder was probably the first to write about the hydrological role of forests. In his *Natural History*, written in the first century, he observed the impact of deforestation on spring flow and rainfall. On the history of this idea, see Vazken Andreassian, "Waters and Forests: From Historical Controversy to Scientific Debate," *Journal of Hydrology* 291, no. 1 (2004): 2.

16. In 1864, the American protoconservationist and watershed management advocate George Perkins Marsh suggested "it is well established" that forests protect spring flows. See George Perkins Marsh, *Man and Nature* (Seattle: University of Washington Press, [1864] 2003), 171; and David Lowenthal, *George Perkins Marsh: Prophet of Conservation* (Seattle: University of Washington Press, 2000).

17. Joseph Kittredge, *Forest Influences* (New York: McGraw-Hill, 1948).

18. On the first paired watershed experiment, see C. G. Bates and A. J. Henry, "Forest and Streamflow Experiment at Wagon Wheel Gap, Colorado," *Monthly Weather Review Supplement* 3 (Washington, DC: USDA Weather Bureau, 1928); On the precursors to and development of forest hydrology, see McCulloch and Robinson, "History of Forest Hydrology."

19. Todd Shallat, *Structures in the Stream: Water, Science, and the Rise of the U.S. Army Corps of Engineers* (Austin: University of Texas Press, 1994), 202.

20. Ibid., 8, 203.

21. On the conflict between the United States Forest Service and the Army Corps of Engineers, see Gordon B. Dodds, "The Stream-Flow Controversy: A Conservation Turning Point," *The Journal of American History* 56, no. 1 (1969): 59–69. Dodds shows how friction between foresters and engineers turned on the efficacy of watershed forests as regulators of stream flow and flooding, framed in terms of forests' "sponge effect." According to Dodds, the Army Corps publicly critiqued the foresters' arguments for basinwide water management, which threatened civil engineering's hegemony over navigation and flood control. In the strongest critique, Army Corps chief H. M. Chittenden argued that foresters' claims had feeble empirical underpinnings.

22. George Goethals, chief engineer of the canal when it was opened and first civil governor of the Canal Zone in 1914, was a graduate of West Point, the Corps' intellectual center, and had managed a number of domestic projects for the Corps before coming to Panama. See Shallat, *Structures in the Stream*, 79–116.

23. Pre-1970s research on Panamanian forests by foreign scientists includes: Paul Allen, "The Timber Woods of Panama," *Ceiba* 10, no. 1 (1964): 17–61; Laurence J. Cummings, *Forestry in Panama* (Panama City: SICAP, 1956); L. R. Holdridge and Gerardo Budowski, "Report on an Ecological Survey of the Republic of Panama," *Caribbean Forester* 17 (1956): 92–110; Henri Pittier, "Our Present Knowledge of the Forest Formations of the Isthmus of Panama," *Journal of Forestry* 16, no. 1 (1918): 76–84. For an overview of forestry education in Latin America before 1960, which does not list any programs in Panama, see Gerardo Budowski, "Forestry Training in Latin America," *Caribbean Forester* (January–June 1961): 33–38.

24. In 1967, the Panamanian Agrarian Reform Commission and US Agency for International Development worked together to relocate three hundred landless families living in "*areas de peligro*" (danger areas) in the Canal Zone to Panamanian territory near Madden Lake. See Government of Panama (hereafter GOP), *Memoria Ministerio de Planificaccion y Politica Economia* (1970), 78–79. The governments' selection of the Madden area as a resettlement site suggests that canal water concerns were not central to the squatter problem at this time, because the families were moved into a future site of watershed management.

25. Negotiation and Planning Records for 1977 Treaty, Squatters and Farming in Canal Zone Lands, 1965–1966, National Archives at College Park, Maryland (hereafter NACP) RG 185, Entry 99, Box 2, File C/REP 7/1.

26. Working Paper for the Panama Canal Review Committee Principles, June 3, 1974, Subject: The Problem of Panamanian Squatters and Territorial Encroachments in the Canal Zone, NACP RG 185, Entry 98, Box 25, File C/REP 7/2, Part 1.

27. Panama Review Committee, Meeting of July 31, 1974, 387th Meeting—C/74–360; Memorandum from the governor to executive secretary, Aug. 21, 1974, NACP RG 185, Entry 98, Box 25, File C/REP 7/2, Part 1.

28. Memo, US Embassy, January 1975, NACP RG 185, Entry 98, Box 25, File C/REP 7/2, Part 1.

29. Panama Review Committee, Meeting of August 6, 1975, 402d Meeting— —C/75–266, NACP RG 185, Entry 98, Box 25, File C/REP 7/2, Part 1.

30. For detailed histories of the treaty and negotiations, see Walter LaFeber, *The Panama Canal: The Crisis in Historical Perspective* (New York: Oxford University Press, 1978), 160–216; and John Major, *Prize Possession: The United States and the Panama Canal, 1903–1979* (Cambridge: Cambridge University Press, 1993), 329–360. Article VI of the Panama Canal Treaties of 1977 was dedicated specifically to the "Protection of the Environment."

31. Panama Canal Treaties of 1977, Article VII, 3, of the Agreement in Implementation of Article III.

32. Subcommittee on the Panama Canal, 1980, Panama Canal Environmental Issues Oversight, 128.

33. Author interview with Stanley Heckadon-Moreno, October 8, 2009. Transnational networks of environmental expertise are documented in the annual reports of Panama's Ministry of Agriculture and natural resource agency (RENARE) throughout the 1970s and 1980s. I also conducted interviews with RENARE staff from this period who supported Heckadon-Moreno's claims.

34. Wadsworth, "Deforestation: Death to the Panama Canal," 24.

35. Curtis Larson, "Erosion and Sediment Yields as Affected by Land Use and Slope in the Panama Canal Watershed," in *Third World Congress on Water Resources*, Vol. 3–4 (Mexico City: International Water Resources Association, 1979), 1086–1095.

36. Evidence of early watershed management work is scattered across government annual reports. See GOP, *Memoria Ministerio de Desarrollo Agropecuario* (1973), 330; GOP, *Memoria Ministerio de Desarrollo Agropecuario* (1975), 269; GOP, *Memoria Ministerio de Desarrollo Agropecuario* (1976), 158. MIDA employees also received training from international organizations. For example, the Inter-American Institute for Cooperation on Agriculture provided a watershed management course in 1974 and 1975. See GOP, *Memoria Ministerio de Desarrollo Agropecuario* (1975), 275, 310. Early watershed research appeared in Cesar Isaza and Blas Moran, "Importancia Del Manejo de La Cuenca Del Canal de Panama" (n.p.: GOP, 1976); and Cesar Isaza and Blas Moran, "National Development and Recovery of the Canal Zone" (n.p.: GOP, 1978).

37. ROCAP-USAID/Panama, "Joint Evaluation of the Watershed Management Project #525-T-049" (1981), 6.

38. T. C. Smith, "The Drainage Basin as an Historical Unit for Human Activity," in *Introduction to Geographical Hydrology*, ed. R. J. Chorley (London: Methuen, 1971), 20.

39. "Watershed," Oxford English Dictionary.

40. Bruce Hooper, *Integrated River Basin Governance: Learning from International Experience* (London: IWA Publishing, 2005), 24.

41. McCulloch and Robinson, "History of Forest Hydrology," 193.

42. Lowenthal, *George Perkins Marsh*, 281.

43. John Wesley Powell, "Institutions for the Arid Lands," *The Century Magazine* xl, no. 1 (May 1890): 111.

44. Teclaff, *The River Basin in History and Law*, 52.

45. On the global dissemination of the TVA model after the Second World War, see, Smith, "The Drainage Basin as an Historical Unit for Human Activity," 28; W. M. Adams, *Green Development: Environment and Sustainability in the Third World.* (London: Routledge, 1990), 218–219. For a poststructuralist critique of development, see Arturo Escobar, *Encountering Development: The Making and Unmaking of the Third World* (Princeton, NJ: Princeton University Press, 1995).

46. David Lilienthal, the TVA's former codirector and chairman, visited the Colombia to advise the year it was established. See Alberto Patiño Mejía, "La Corporación del Valle del Cauca Promotora de Desarrollo Rural," in *La Cuenca Del Canal de Panama: Actas de Los Seminarios-Talleres*, ed. Stanley Heckadon-Moreno (n.p.: GOP, 1986), 261.

47. Teclaff, *The River Basin in History and Law*, 138–139.

48. The presentations given at these meetings are published in *La Cuenca Del Canal de Panama: Actas de Los Seminarios-Talleres*, ed. Stanley Heckadon-Moreno (n.p.: GOP, 1986).

49. Author interview with Stanley Heckadon-Moreno, October 15, 2009.

50. Patiño Mejía, "La Corporación del Valle del Cauca Promotora del Desarrollo Rural," 259–272.

51. ROCAP-USAID/Panama, 1981, 6.

52. Jane Wolff, "Redefining Landscape," in *The Tennessee Valley Authority: Design and Persuasion*, ed. Tim Culvahouse (New York: Princeton Architectural Press, 2007), 54.

53. Huxley, *TVA: Adventure in Planning*, quoted in Wolff, "Redefining Landscape," 56.

54. Wolff, "Redefining Landscape," 57, 63.

55. Paul Edwards, *A Vast Machine* (Cambridge, MA: MIT Press, 2010), 12.

56. The concept of moral economy, or the set of cultural norms and expectations concerning the legitimate social roles of particular groups within the economy, was first developed in E. P. Thompson, "The Moral Economy of the English Crowd in the Eighteenth Century," *Past and Present* 50, no. 1 (1971): 76–136.

57. Author interview, October 12, 2008.

58. The seminal work on enrollment and translation in actor network theory is Michel Callon, "Some Elements of a Sociology of Translation: Domestication of the Scallops and the Fishermen of St. Brieuc Bay," in *Power, Action and Belief: A New Sociology of Knowledge?*, ed. John Law (London: Routledge, 1986), 196–223.

59. ROCAP-USAID/Panama, 1981, 7.

60. Rosa María Cortéz, "Características Generales de La Población," in *La Cuenca del Canal de Panama: Actas de los Seminarios-Talleres*, ed. Stanley Heckadon-Moreno (n.p.: GOP, 1986), 45.

61. Hugh Raffles, "The Uses of Butterflies," *American Ethnologist* 28, no. 3 (2001): 513.

62. Author, personal communication with Francisco Herrera (Panamanian anthropologist), June 20, 2013.

63. Author, personal communication with Francisco Herrera, June 20, 2013. See Luis Pinzon and Jose Esturain, "Vigilancia de Los Bosques," in *La Cuenca del Canal de Panama: Actas de los Seminarios-Talleres*, ed. Stanley Heckadon-Moreno (n.p.: GOP, 1986), 213–214.

64. Pinzon and Esturain, "Vigilancia de Los Bosques," 10.

65. On the *long durée* of Panamanian agriculture, see Charles Bennett, *Human Influences on the Zoogeography of Panama* (Berkeley: University of California Press, 1968). For a historical overview of environmental change in the Chagres River basin, see Stanley Heckadon-Moreno, "Light and Shadows in the Management of the Panama Canal Watershed," in *The Rio Chagres: A Multidisciplinary Profile of a Tropical River Basin*, ed. Russell S. Harmon (New York: Kluwer Academic/Plenum Publishing, 2005), 37.

66. USAID, "Evaluation of USAID's Strategic Objective for the Panama Canal Watershed 2000–2005" (June 2005); prepared by Development Alternatives, Inc., http://pdf.usaid.gov/pdf_docs/Pdacj404.pdf

67. Piers Blaikie and Harold Brookfield, "Defining and Debating the Problem," in *Land Degradation and Society* (London: Metheun, 1987), 6–7.

Chapter 4

1. Alejandro Balaguer, "Cuenca Milagrosa / The Miracle of the Watershed," *Escapes Panama: Magazine of Air Panama* (January–February 2009): 44–45.

2. Frank Wadsworth, "Deforestation: Death to the Panama Canal," in *US Strategy Conference on Tropical Deforestation* (Washington, DC: US Department of State and US Agency for International Development, 1978), 22.

3. On maps as spatial propositions that bring territory into existence, see Denis Wood, *Rethinking the Power of Maps* (New York: Guilford, 2010), 39–64.

4. "An Interview with Frank Robinson, of the Meteorological and Hydrographic Branch, July 7, 1982," 1; National Archives at College Park, Maryland, RG 185, Entry 98, 007, Box 1, Records Relating to the History of the Implementation of the 1977 Panama Canal Treaty, 1968–85.

5. "An Interview with Frank Robinson," 11.

6. Author interview with "Robbie" Robinson, September 25, 2009.

7. "An Interview with Frank Robinson," 17.

8. Quoted in Jan Meriwether, "Canal Water Watchers," *Panama Canal Review*, October 1, 1981, 13.

9. Ibid., 14.

10. See the discussion of boundary objects and people that cross social worlds in Susan Leigh Star and James R. Griesemer, "Institutional Ecology, 'Translations' and Boundary Objects: Amateurs and Professionals in Berkeley's Museum of Vertebrate Zoology, 1907–39," *Social Studies of Science* 19, no. 3 (1989): 387–420

11. Meriwether, "Canal Water Watchers," 14–15.

12. Paul Simons, "Nobody Loves a Canal with No Water," *New Scientist* (October 1989): 49.

13. "An Interview with Frank Robinson," 11.

14. Author interview with Luis Alvarado, October 7, 2009.

15. On this philosophy of history, see Bruno Latour, "The Historicity of Things: Where Were Microbes Before Pasteur?," in *Pandora's Hope: Essays on the Reality of Science Studies* (Cambridge, MA: Harvard University Press, 1999), 145–173.

16. Ibid., 168.

17. Frank H. Robinson, *A Report on the Panama Canal Rain Forests* (n.p.: Panama Canal Commission, 1985), 41–43.

18. Ibid., 3.

19. See Henri Lefebvre, *The Production of Space* (Oxford: Blackwell, [1974] 1991), 26.

20. Gregory Bateson, *Steps to an Ecology of Mind* (Chicago: University of Chicago Press, [1972] 2000), 479.

Chapter 5

1. Bonifacio Pereira Jiménez, *Biografía del Río Chagres* (Panama City: Imprenta Nacional, 1964).

2. Peter A. Szok, *"La Última Gaviota": Liberalism and Nostalgia in Early Twentieth-Century Panama* (Westport, CT: Greenwood Publishing Group, 2001), 38.

3. Frederick Upham Adams, *Conquest of the Tropics: The Story of the Creative Enterprises Conducted by the United Fruit Company* (Garden City, NY: Doubleday, Page and Company, 1914), 38.

4. Julie Greene, *The Canal Builders: Making America's Empire at the Panama Canal* (New York: Penguin, 2009), 23.

5. Joseph Masco, *The Nuclear Borderlands: The Manhattan Project in Post-Cold War New Mexico* (Princeton, NJ: Princeton University Press, 2006), 14.

6. The argument, whether explicit or implicit, that the isthmus's geographic position has been the prime mover of national history is a powerful discourse in Panamanian scholarship. For a critique of "natural advantages" discourse in the history of Chicago, see William Cronon, *Nature's Metropolis: Chicago and the Great West* (New York: W. W. Norton), 55–96. On the role of physical geography and historical inertia in shaping the spatial structure of transportation networks, see Jean-Paul Rodrigue, Claude Comtois, and Brian Slack, *The Geography of Transport Systems* (London: Routledge, 2013).

7. This geological history is drawn from the work of Anthony Coates and colleagues. Geologists debate the timing and direction of the complex tectonic plate movements that led to the formation of modern Central America and Panama. This is a summary of the most widely accepted findings. See Anthony G. Coates, "The Forging of Central America," in *Central America: A Natural and Cultural History*, ed. Anthony G. Coates (New Haven, CT: Yale University Press, 1997), 1–37; Anthony G. Coates, Jeremy B. C. Jackson, Laurel S. Collins, Thomas M. Cronin, Harry J. Dowsett, Laurel M. Bybell, Peter Jung, and Jorge A. Obando, "Closure of the Isthmus of Panama: The Near-Shore Marine Record of Costa Rica and Western Panama," *Geological Society of American Bulletin* 104 (1992): 814–828; Laurel S. Collins, Anthony G. Coates, William A. Berggren, Marie Pierre Aubry, and Jihun Zhang, "The Late Miocene Panama Isthmian Strait," *Geology* 24, no. 8 (2006): 687–690.

8. Richard Cooke, Dolores Piperno, Anthony J. Ranere, Karen Clary, Patricia Hansell, Storrs Olson, Valerio L. Wilson, and Doris Weiland, "La Influencia de las Poblaciones Humanas Sobre los Ambientes Terrestres de Panama Entre el 10,000 A.C. y el 500 D.C.," in *Agonia de La Naturaleza*, ed. Stanley Heckadon-Moreno and Jaime Espinosa Gonzalez (Panama City: IDIAP and STRI, 1985), 3–25; Richard Cooke, "The Native Peoples of Central America during Precolumbian and Colonial Times," in *Central America: A Natural and Cultural History*, ed. Anthony G. Coates (New Haven, CT: Yale University Press, 1999), 145.

9. On the contentious debates about precontact population estimates and their environmental consequences, see Charles C. Mann, *1491: New Revelations of the Americas before Columbus* (New York: Vintage Books, 2006). For a key work related to the upward revision of population numbers, see William Denevan, "The Pristine Myth: The Landscape of the Americas in 1492," *Annals of the Association of American Geographers* 82, no. 3 (1992): 369–385.

10. Scholars debate the degree to which indigenous peoples modified the isthmian environment before contact with Europeans. Some, like geographer Charles Bennett, argue that landscape modifications were significant and that the colonial era was not the beginning of environmental decline, but rather an "ecological retreat of man" that lasted until 1900. See Charles Bennett, *Human Influences on the Zoogeography of Panama* (Berkeley: University of California Press, 1968), 55. Other scholars suggest that indigenous Panamanian populations were small and that their impacts were minimal. See Stanley Heckadon-Moreno, "Light and Shadows in the Management of the Panama Canal Watershed," in *The Rio Chagres: A Multidisciplinary Profile of a Tropical River Basin*, ed. Russell S. Harmon (New York: Kluwer Academic/ Plenum Publishing, 2005), 31; and Omar Jaén Suárez, *Hombres y Ecología En Panamá* (Panama City: Editorial Universitaria, Smithsonian Tropical Research Institute, 1981).

11. Gerstel Mack, *The Land Divided: A History of the Panama Canal and Other Isthmian Canal Projects* (New York: Knopf, 1944), 3–9.

12. Lancelot S. Lewis, *The West Indian in Panama: Black Labor in Panama, 1850–1914* (Washington, DC: University Press of America, 1980), 5.

13. Mack, *The Land Divided*, 41.

14. Quoted in ibid., 53.

15. Ibid., 53–55.

16. Alfredo Castillero Calvo, "Transistmo y Dependencia: El Caso del Istmo de Panamá," *Loteria* 211 (1973): 25–56.

17. Mack, *The Land Divided*, 41.

18. Lewis, *The West Indian in Panama*, 5.

19. Aims McGuinness, *Path of Empire: Panama and the California Gold Rush* (Ithaca, NY: Cornell University Press, 2008), 18.

20. Ibid., 20.

21. Walter LaFeber, *The Panama Canal* (New York: Oxford University Press, 1990), 34.

22. McGuinness, *Path of Empire*, 32.

23. Ibid.

24. Ibid., 35–36.

25. F. N. Otis, *Illustrated History of the Panama Railroad* (Pasadena, CA: Socio-Technical Books, [1861] 1971), 25.

26. McGuinness, *Path of Empire*, 54–57.

27. McCullough, *The Path Between the Seas*, 35.

28. Ibid., 37.

29. Castillero Calvo, "Transistmo y Dependencia," n.p.

30. Donald Duke, "Forward to Illustrated History of the Panama Railroad," in Otis, *Illustrated History of the Panama Railroad*, 1.

31. Guillermo Castro Herrera, "Pro Mundi Beneficio: Elementos Para Una Historia Ambiental de Panamá," *Revista Tareas* 120 (May–August 2005): 83; Jaén Suárez, *Hombres y Ecología en Panamá*.

32. Castillero Calvo, "Transistmo y Dependencia," n.p.

33. Adams, *Conquest of the Tropics*, 35.

34. Ibid., 38.

35. Lewis, *The West Indian in Panama*, 19–20. See also Gustave Anguizola, "Negroes in the Building of the Panama Canal," *Phylon* 29, no. 4 (1968): 352.

36. Otis, *Illustrated History of the Panama Railroad*, 81.

37. Matthew Parker, *Panama Fever: The Battle to Build the Panama Canal* (London: Hutchinson, 2009), 48–55; and McCullough, *The Path Between the Seas*, 287.

38. Parker, *Panama Fever*, 59–63.

39. Ibid., 64–65.

40. Quoted in Parker, *Panama Fever*, 63–64.

41. Mack, *The Land Divided*, 317.

42. Parker, *Panama Fever*, 91.

43. Mack, *The Land Divided*, 338–341; Parker, *Panama Fever*, 92–93.

44. Michael L. Conniff, *Black Labor on a White Canal* (Pittsburgh, PA: University of Pittsburgh Press, 1985), 17.

45. Mack, *The Land Divided*, 329–331.

46. Any mortality estimates are only educated guesses, but this number is widely cited, as discussed in McCullough, *The Path Between the Seas*, 235; and Conniff, *Black Labor on a White Canal*, 20.

47. Parker, *Panama Fever*, 174.

48. Ibid., 186.

49. Ibid., 185–193.

50. John Major, *Prize Possession: The United States and the Panama Canal, 1903–1979* (Cambridge: Cambridge University Press, 1993), 37.

51. LaFeber, *The Panama Canal*, 23.

52. Greene, *The Canal Builders*, 21.

53. Conniff, *Black Labor on a White Canal*, 17.

54. Jaén Suárez, *Hombres y Ecología en Panamá*, 126.

55. George E. Roberts, *Investigación Económica de La República de Panamá* (Managua: Fundación UNO, [1932] 2006), 200.

56. This claim is also supported by oral histories from Gatun Lake published in Stanley Heckadon-Moreno, *Los Sistemas de Producción Campesinos y Los Recursos Naturales En La Cuenca Del Canal* (n.p.: Government of Panama, 1981), 6–12.

57. Alfredo Castillero Calvo, *La Ruta Transístmica y Las Comunicaciones Marítimas Hispanas Siglos XVI a XIX* (Panama City: Ediciones Nari, 1984).

58. Castillero Calvo, "Transistmo y Dependencia."

Chapter 6

1. This approach is inspired by geographer Karen Bakker's work on water politics and governance. See Karen Bakker, "Water: Political, Biopolitical, Material," *Social Studies of Science* 42, no. 4 (2012): 618.

2. Mark L. Gillem, *America Town: Building the Outposts of Empire* (Minneapolis: University of Minnesota Press, 2007), 3–4.

3. Hay–Bunau-Varilla Treaty (Panama Canal Treaty) 1903, Articles II and III.

4. Walter LaFeber, *The Panama Canal: The Crisis in Historical Perspective* (New York: Oxford University Press, 1978), 37–38; and John Major, *Prize Possession: The United States and the Panama Canal, 1903–1979* (Cambridge: Cambridge University Press, 1993), 47.

5. For a summary of the land that the United States expropriated from Panama between 1908 and 1931, see John Lindsay-Poland, *Emperors in the Jungle: The Hidden History of the US in Panama* (Durham, NC: Duke University Press, 2003), 28.

6. Julie Greene, *The Canal Builders: Making America's Empire at the Panama Canal* (New York: Penguin, 2009), 23, 192–194.

7. Ibid., 57–59.

8. David McCullough, *The Path Between the Seas: The Creation of the Panama Canal, 1870–1914* (New York: Simon and Schuster, 2001), 529–554.

9. Greene, *The Canal Builders*, 44–45.

10. McCullough, *The Path Between the Seas*, 547–548.

11. Greene, *The Canal Builders*, 136.

12. McCullough, *The Path Between the Seas*, 545.

13. E. A. Goldman to Edward William Nelson, March 16, 1911, Smithsonian Institution Archives, RU 7364, Box 4, Folder 13, Edward William Nelson and Edward Alphonso Goldman Collection, circa 1873–1946 and undated.

14. McCullough, *The Path Between the Seas*, 543.

15. Ibid., 554.

16. Scott Kirsch and Don Mitchell, "Earth-Moving as the 'Measure of Man': Edward Teller, Geographical Engineering, and the Matter of Progress," *Social Text* 54, no. 16 (1998): 100–134.

17. *Report of the Board of Consulting Engineers, Panama Canal* (Washington, DC: Government Printing Office, 1906)

18. Ibid., xiv, xvii.

19. *Report of Board of Consulting Engineers, Panama Canal.*

20. Quoted in McCullough, *The Path Between the Seas*, 483.

21. "Canals New and Old," *Chicago News*, May 4, 1908.

22. *Report of Board of Consulting Engineers, Panama Canal*, xvi.

23. Ibid., 79–80.

24. Reuben E. Bakenhus, Harry S. Knapp, and Emory R. Johnson, *The Panama Canal: Comprising Its History and Construction, and Its Relation to the Navy, International Law and Commerce* (New York: John Wiley and Sons, 1915), 81.

25. *Report of Board of Consulting Engineers, Panama Canal*, 5.

26. "The Width of the Locks," *Star and Herald*, February 10, 1908.

27. The role of technical standards in transportation is a rich area of inquiry in the history of technology. Scholars have shown that the establishment of standards across systems that require coupling tend to create strong path dependence. See Lawrence Busch, *Standards: Recipes for Reality* (Cambridge, MA: MIT Press, 2011), 64; and Michel Callon, "Techno-Economic Networks and Irreversibility," in *A Sociology of Monsters: Essays on Power, Technology and Domination*, ed. John Law (London: Routledge, 1995), 159. When transportation systems—or any technical systems—are decentralized and many actors must agree to change the standard in question, negotiation often becomes difficult. For example, standardized railroad gauges were first developed by British Engineer George Stephenson in 1830, but engineers around the world developed many different, incompatible track gauges because they did not anticipate that railroads would replace inland water transportation. See Busch, *Standards*, 64. As railroad systems were coupled and their gauges standardized during the second half of the nineteenth century—the outcome of state intervention in the United States—technical standardization precipitated social standardization. For example, as Bill Cronon shows in his history of Chicago, the schedules of newly coupled railroad systems standardized, for the first time, how people across a large area experienced time and seasons. See Bill Cronon, *Nature's Metropolis: Chicago and the Great West* (New York: W. W. Norton and Company, 1991), 76.

28. Most water flows out of the canal system via the locks, but it also exits through the Gatun Dam spillway and hydroelectric turbine, a system that diverts it for consumption in the terminus cities, and evaporation.

29. George Matthew, "Chagres River and Gatun Lake Watershed Hydrological Data, 1907–1948" (n.p.: Panama Canal Section of Meteorology and Hydrology, 1949), 3.

30. Henry Abbot, *Problems of the Panama Canal* (New York: The Macmillan Company, 1905), 105.

31. Frank Gause and Charles C. Carr, *The Story of Panama: The New Route to India* (Boston: Silver, Burdett and Company, 1912), 62–63.

32. The earliest reported Chagres River basin survey was published as a map in 1864 by the Colombian government. In 1873, an American expedition led by Commander E. P. Lull and Lieutenant F. Collins, both of the US Navy, made a traverse survey along the Chagres River. There were more than a hundred in the party and they ran line levels, prepared maps, charts, and statistical tables. The Isthmian Canal

Commission conducted a rough watershed survey between 1904 and 1907. See McCullough, *The Path Between the Seas*, 63.

33. C. M. Saville, assistant engineer, to George Goethals, chief engineer, January 22, 1909, National Archives at College Park, Maryland, RG 185, Entry 30, File 67.c.61(2), Chagres River Watershed; Surveys, etc.–General.

34. "Chagres River Topography—Survey Begun on the Basin Above Gamboa," *Canal Record*, November 18, 1908.

35. George Goethals, chief engineer, to Luke Wright, secretary of war, December 25, 1908, NACP RG 185, Entry 30, File 33.d.37(1), Lands of Canal Zone—Agricultural Possibilities.

36. Quoted in Leo S. Rowe, "The Work of the Joint International Commission on Panamanian Land Claims," *The American Journal of International Law* 8, no. 4 (1914): 738–757.

37. January 27, 1928, NACP RG 185, Entry 30, File 33.b.51(1).

38. "Removal of Gatun Village," *Canal Record*, April 8, 1908.

39. Greene, *The Canal Builders*, 151.

40. "The Depopulation of the Canal Zone," *Canal Record*, March 21, 1917.

41. This shift is documented throughout NACP RG 185, Entry 30, File 33.b.51 (1 and 2).

42. Quoted in Rowe, "The Work of the Joint International Commission on Panamanian Land Claims," 739.

43. NACP RG 185, Entry 30, File 33.d.11/g(2), Boundary Markers.

44. "Emigrants from the Lake Area," *Canal Record*, September 10, 1913, 22.

45. "Back to Zone for Cultivation," *The Workman*, 1921, December 17.

46. "Extension of Control Over Lake Area," *Canal Record*, May 29, 1912.

47. "Emigrants from the Lake Area," *Canal Record*, September 10, 1913, 22.

48. "The Depopulation of the Canal Zone," *Canal Record*, March 21, 1917.

49. Rowe, "The Work of the Joint International Commission on Panamanian Land Claims."

50. N. A. Becker, acting land agent, to Jay Morrow, governor of the Canal Zone, June 18, 1923, NACP RG 185, Entry 30, File 33.b.51(5), Licensing Land in Canal Zone for Agricultural Purposes.

51. "Hamlets and Towns Whose Sites Will Be Covered by Gatun Lake," *Canal Record*, December 6, 1911.

52. "Malicious Interference with Boundary Marks," *Canal Record*, August 14, 1912.

53. Letter by counsel of Panama Railroad, Canal Zone land agent, and chief of Police and Fire Division to C. A. McIlvaine, executive secretary, Nov. 24, 1923, NACP RG 185, Entry 30, File 33.b.51(5).

54. Benigno Palma, "Vox climantis in deserto," *The Independent*, November 14, 1914.

55. US Library of Congress, Manuscript Division, Panama Collection of the Canal Zone Library Museum, Box 32.

56. Stanley Heckadon-Moreno, "Light and Shadows in the Management of the Panama Canal Watershed," in *The Rio Chagres: A Multidisciplinary Profile of a Tropical River Basin*, ed. Russell S. Harmon (New York: Kluwer Academic/Plenum Publishing, 2005), 33.

57. R. Z. Kirkpatrick, "Madden Dam Will Insure Water Supply for Gatun Lake Development," *Iowa Engineer* (March 1934): 84.

58. On the role of canal construction in creating environmental conditions that facilitated the spread of mosquito-borne disease, see Lindsay-Poland, *Emperors in the Jungle*, 32; and Paul S. Sutter, "Nature's Agents or Agents of Empire?: Entomological Workers and Environmental Change During the Construction of the Panama Canal," *Isis* 98, no. 4 (2007): 740–746. The quote attributed to William Gorgas's wife is published in Lindsay-Poland, *Emperors in the Jungle*, 32.

59. Alberto E. Calvo Sucre, *Análisis Histórico del Desarrollo de la Salud Publica en la Republica de Panamá, 1903-2006* (Panama City: Editorial Universitaria Carlos Manuel Gasteazoro, 2007), 45–47; Sutter, "Nature's Agents or Agents of Empire?," 740–741; and Paul S. Sutter, "Tropical Conquest and the Rise of the Environmental Management State: The Case of US Sanitary Efforts in Panama," in *Colonial Crucible: Empire in the Making of the Modern American State*, ed. Alfred W. McCoy and Francisco A. Scarano (Madison: University of Wisconsin Press, 2009), 317–326.

60. On infrastructures and citizenship, see Madeleine Akrich, "The De-Scription of Technical Objects," in *Shaping Technology/Building Society: Studies in Sociotechnical Change*, ed. Wiebe E. Bijker and John Law (Cambridge, MA: MIT Press, 1992), 205–224; For recent anthropological work, see Nikhil Anand, "Pressure: The Politechnics of Water Supply in Mumbai," *Cultural Anthropology* 26, no. 4 (2011): 542–564 and Antina von Schnitzler, "Citizenship Prepaid: Water, Calculability, and Techno-Politics in South Africa," *Journal of Southern African Studies* 34, no. 4 (2008): 899–917.

61. "Canal Zone Lands," *Canal Record*, Aug. 2, 1911.

Chapter 7

1. Government of Panama, 2000, Census. The national census of 2000 recorded 574 residents in Limón, compared to 533 residents in 1980.

2. See the section of chapter 1 on the consolidation of global transportation infrastructure, particularly the discussion of how the rise of intermodal transportation around the shipping container reduced the size of labor forces required at "breaks" between transportation systems and modes like ships and railroads. Over the last several decades, ports became more mechanized, meaning that fewer, but more skilled workers were needed than was previously the case.

3. Gil Blas Tejeira, *Pueblos Perdidos* (Panama City: Editorial Universitaria, [1962] 2003), 99.

4. Alessandro Portelli, *The Death of Luigi Trastulli and Other Stories: Form and Meaning in Oral History* (Albany: State University of New York Press, 1991), 1.

5. Tejeira, *Pueblos Perdidos*, 157.

6. Author interview with resident of Limón, June 28, 2008.

7. Stanley Heckadon-Moreno, *Los Sistemas de Producción Campesinos y Los Recursos Naturales En La Cuenca Del Canal* (n.p.: Government of Panama, 1981), 5.

8. James Ferguson, *Expectations of Modernity: Myths and Meanings of Urban Life on the Zambian Copperbelt* (Berkeley: University of California Press, 2006), 238.

9. The idea that shifts in infrastructural organization produce "orphans," which can be understood as individuals, groups, or practices, comes from Paul Edwards, Steven J. Jackson, Geoffrey C. Bowker, and Cory P. Knobel, *Understanding Infrastructure: Dynamics, Tensions, and Design/Report of a Workshop on History and Theory of Infrastructure: Lessons for New Scientific Cyberinfrastructures* (Ann Arbor: University of Michigan, 2007), 25–27.

Chapter 8

1. Otis W. Barrett "Horticulture in Canal Zone," draft of an article for *The Standard Cyclopedia of Horticulture* by L.H. Bailey (n.p., 1915); see National Archives at College Park, Maryland (hereafter NACP) RG 185, Entry 34, File 33.h.3, Horticultural Work on the Isthmus.

2. Hugh Bennett and William Taylor, *The Agricultural Possibilities of the Canal Zone* (Washington, DC: Government Printing Office, 1912), 39.

3. Geographer Stephen Frenkel characterizes the establishment of the Canal Zone as a suburban residential landscape between 1912 and 1940 as an effort to separate

white North Americans from an array of human and nonhuman "others," namely the Panamanian "jungle," Panamanian cities, and West Indian laborers and Spanish-speaking Panamanians; Stephen Frenkel, "Geographical Representations of the 'Other': The Landscape of the Panama Canal Zone," *Journal of Historical Geography* 28, no. 1 (2002): 85–99.

4. "Banana Shipments from Canal Zone Ports," *Canal Record*, January 18, 1928; Honduras, the top Central American producer, exported 29 million bunches in 1929. See John Soluri, *Banana Cultures: Agriculture, Consumption, and Environmental Change in Honduras and the United States* (Austin: University of Texas Press, 2005), 52.

5. Staffen Müller-Wille, "Introduction," in *Musa Cliffortiana: Clifford's Banana Plant*, by Carl Linnaeus (Ruggell, Liechtenstein: A.R.G. Ganternet Verlag K.G., 2007), 15–67.

6. The history of the banana's introduction to the Americas is debated, with many contradictory statements in the primary and secondary literature. For more information on these debates, see Robert Langdon, "The Banana as a Key to Early American and Polynesian History," *The Journal of Pacific History* 28, no. 1 (1993): 15–35. For the most common and perhaps the earliest account, which is referenced here, see Gonzalo Fernández de Oviedo, *Natural History of the West Indies* (Chapel Hill, NC: The Orange Printshop, [1526] 1959).

7. Alfred Crosby, *Ecological Imperialism: The Biological Expansion of Europe, 900–1900* (Cambridge: Cambridge University Press, 1986).

8. Carl Linnaeus wrote that Barbadian sugar plantation slaves were provided one to two bananas weekly. In a 2007 introduction to the work, Müller-Wille discusses the relationship among bananas, slavery, and the colonial plantation economy. He writes, "[Linnaeus] overlooked in a typically Eurocentric and supremacist way the degree to which the plantation system depended on a well-developed subsistence and market economy among slaves." See Müller-Wille, "Introduction," in *Musa Cliffortiana: Clifford's Banana Plant*, 28–29.

9. Soluri, *Banana Cultures*, 226.

10. Because attracting local labor was a perpetual problem in Costa Rica, Minor Keith recruited migrant West Indian laborers from Jamaica, where the economy was terrible. On this history, see Aviva Chomsky, *West Indian Workers and the United Fruit Company in Costa Rica, 1870–1940* (Baton Rouge: Louisiana State University Press, 1996), 27; and Steve Marquardt, "'Green Havoc': Panama Disease, Environmental Change, and Labor Process in the Central American Banana Industry," *The American Historical Review* 106, no. 1 (2001): 53–55.

11. Dan Koeppel, *Banana: The Fate of the Fruit That Changed the World* (New York: Hudson Street Press, 2008), 58–60; and Marquardt, "Green Havoc," 55.

12. The early history of the banana-railroad union in Central America is described in Charles David Kepner Jr. and Jay Henry Soothill, *The Banana Empire: A Case Study of Economic Imperialism* (New York: The Vanguard Press, 1935), 155. Bananas have been a topic of a raft of recent works. For academic accounts, see, for example, Philippe Bourgois, *Ethnicity at Work: Divided Labor on a Central American Banana Plantation* (Baltimore, MD: The Johns Hopkins University Press, 1989); Lawrence Grossman, "The Political Ecology of Banana Exports and Local Food Production in St. Vincent, Eastern Caribbean," *Annals of the Association of American Geographers* 83, no. 2 (1993): 347–367; Steve Striffler and Mark Moberg, *Banana Wars: Power, Production, and History in the Americas* (Durham, NC: Duke University Press, 2003); and Karla Slocum, *Free Trade and Freedom: Neoliberalism, Place, and Nation in the Caribbean* (Ann Arbor: University of Michigan Press, 2006); Soluri, *Banana Cultures*. For popular accounts, see Virginia Scott Jenkins, *Bananas: An American History* (Washington, DC: Smithsonian Institution, 2000); Koeppel, *Banana*; and Julian Roche, *The International Banana Trade* (Boca Raton, FL: CRC Press, 1998).

13. Bennett and Taylor, *The Agricultural Possibilities of the Canal Zone*, 39.

14. Bonham C. Richardson, *Panama Money in Barbados, 1900–1920* (Knoxville: University of Tennessee Press, 1985), 197.

15. J. A. LePrince, chief sanitary inspector, to William Gorgas, chief sanitary officer, Jan. 30 1906, NACP RG 185, Entry 30, File 33.d.37(1), Lands of Canal Zone— Agricultural Possibilities.

16. George Goethals, chief engineer, to Jacob Dickison, secretary of war, June 22, 1909, NACP RG 185, Entry 30, File 33.d.37(1).

17. George Goethals to Major Milson, subsistence officer, June 22, 1909. NACP RG 185, Entry 30, File 33.d.37(1).

18. The classic work on moral economy is by historian E. P. Thompson, who argues that food riots in eighteenth century England should not be explained through "an abbreviated view of economic man." He argues that people participated in food riots not simply because they were hungry, but because they were "informed by the belief that they were defending traditional rights and customs ... supported by the consensus of the community"; E. P. Thompson, "The Moral Economy of the English Crowd in the Eighteenth Century," *Past and Present* 50 (1971): 76–136.

19. David Fairchild, US Department of Agriculture agricultural explorer in charge, had a long-term interest in Canal Zone horticulture. See his extensive correspondence with Zone horticulturalists in NACP RG 185, Entry 30, File 33.h.3; and Entry 34, File 33.h.3, Horticultural Work on the Isthmus.

20. David Fairchild, US Department of Agriculture, to William Gorgas, chief sanitary officer, April 21, 1906, NACP RG 185, Entry 30, File 33.3.

21. Henry Schultz, landscape gardener, to J. A. LePrince, chief sanitary inspector, Sept. 5 1907, NACP RG 185, Entry 30, File 33.h.9(1), Gardens—Operation of by the Panama Canal.

22. See NACP RG 185, Entry 30, File 33.h.9(1).

23. Bennett and Taylor, *The Agricultural Possibilities of the Canal Zone*, 39.

24. Ibid., 11.

25. Ibid., 20.

26. Ibid., 47.

27. The region's historical demography, defined by immigration and emigration, make stable historical definitions of "native" both unstable and problematic.

28. Bennett and Taylor, *The Agricultural Possibilities of the Canal Zone*, 46.

29. George Goethals explained his governmental rationale for not wanting West Indians and Panamanians to populate the rural areas of the Canal Zone in George Goethals, *Government of the Canal Zone* (Princeton, NJ: Princeton University Press, 1915), 60–68. The passage quoted comes from p. 64 of the same work.

30. Ibid., 66–67, 65.

31. Michael L. Conniff, *Black Labor on a White Canal: Panama, 1904–1981* (Pittsburgh, PA: University of Pittsburgh Press, 1985); and Omar Jaén Suárez, *Hombres y Ecología En Panamá* (Panama City: Editorial Universitaria, Smithsonian Tropical Research Institute, 1981).

32. Conniff, *Black Labor on a White Canal*, 46.

33. The Jamaican banana trade was devastated through the cumulative effects of drought, hurricanes, and war during this period; Peter Clegg, *The Caribbean Banana Trade: From Colonialism to Globalization* (New York: Palgrave Macmillan, 2002), 34.

34. Reverend S. Moss Loveridge, honorary chaplain of the Canal Zone Mission Churches, to Chester Harding, governor of the Canal Zone, May 4, 1917, NACP RG 185, Entry 34, File 33.h.9(2).

35. For a discussion of canal worker strikes and protests during this era, see Conniff, *Black Labor on a White Canal*, 52-61. The size of the 1920 strike was reported in *Annual Report of the Governor of the Panama Canal for the Fiscal Year Ended June 30, 1920* (Washington, DC: Government Printing Office, 1920), 27.

36. R. K. Morris, quartermaster, to Jay Morrow, governor of the Canal Zone, September 7, 1921, NACP RG 185, Entry 34, File 33.b.51(2a).

37. Memorandum to General Connor, written by Jay Morrow, governor of the Canal Zone, July 11, 1921, NACP RG 185, Entry 34, File 33.b.51(2a).

38. In 1914, bananas represented 65 percent of Panamanian exports. See William T. Scoullar, *Libro Azul de Panama* (Panama City: Imprenta Nacional, 1917), 65.

39. N. A. Becker, land agent, to engineer of maintenance, May 3, 1932, NACP RG 185, Entry 34, File 33.b.51(8); See also Scoullar, *Libro Azul de Panama*, 65.

40. "Ruminations of an Old Timer," *Panama Times*, May 9, 1926.

41. C. A. McIlvaine, executive secretary, to Constantine Graham, British Charge d'Affairs in Panama, August 16 1922, NACP RG 185, Entry 34, File 33.b.51(4).

42. The extent of preconstruction forest cover is unclear, but there are numerous accounts of the remains of old farms being found within the secondary forest that grew up across the rural Canal Zone after the depopulation order of 1912.

43. Author interview with resident of Salamanca, Panama, November 7, 2008,

44. Marquardt, "Green Havoc," 55.

45. "Bananas," *Panama American*, July 1, 1927.

46. "Banana Exports from Cristobal," *Canal Record*, July 23, 1924.

47. The weekly income figures for smallholder banana farmers come from "Agricultural Experiments," *Star and Herald*, January 23, 1926. I estimated 2014 buying power using the United States Bureau of Labor Statistics inflation calculator, accessed at http://www.bls.gov/data/inflation_calculator.htm. Archival accounts of the lived experiences of banana producers and land lessees are scarce, so I draw here on oral histories that I conducted with them and their descendants. My interviews were conducted during 2008 and 2009 in two Gatun Lake communities that were former hubs of the banana trade, as well as one community upstream from which farmers canoed down to the lake to sell their bananas.

48. Kepner and Soothill, *The Banana Empire*, 266–267.

49. Corrine Browning Feeney, "The Story of the Banana Industry in Panama," *Star and Herald*, June 6, 1925.

50. Temperate zone settlers in Panama persistently commented on the problem of establishing rational economic behavior among peoples that they considered to be unmotivated by capitalist accumulation, or were, as they put it, "lazy." Scholars have examined this problem at the historical (e.g., industrial revolution in Europe) or geographical margins of capitalism (i.e., market integration). As Michael Taussig writes, "The first reaction of such persons to their (usually forced) involvement in modern business enterprises as wage workers is frequently, if not universally, one of indifference to wage incentives and to the rationality that motivates homo oeconomicus. This response has time and time again frustrated capitalist entrepreneurs." Michael Taussig, *The Devil and Commodity Fetishism in South America* (Chapel Hill: The University of North Carolina Press, 1980), 19.

51. Feeney, "The Story of the Banana Industry in Panama."

52. "Shipments of Bananas from Gatun Lake Area," *Canal Record*, November 28, 1923.

53. United Fruit Company general agent in Cristobal, Panama, to Jay Morrow, governor of the Canal Zone, Nov. 17, 1923, NACP RG 185, Entry 34, File 33.h.5(1), Bananas—Culture and Production in the Canal Zone and Vicinity.

54. Marquardt, "Green Havoc," 54.

55. Feeney, "The Story of the Banana Industry in Panama."

56. Philippe Bourgois, "100 Years of United Fruit Company Letters," in *Banana Wars: Power, Production, and History in the Americas*, ed. Steve Strifler and Mark Moberg (Durham, NC: Duke University Press, 2003), 113–114.

57. Panama Canal Zone press release reprinted in *Star and Herald*, May 9, 1923.

58. "Bananas from Gatun Lake Region," *Canal Record*, November 7, 1923.

59. Jaén Suárez, *Hombres y Ecología en Panamá*, 126.

60. "Shipments of Bananas from Cristobal," *Canal Record*, January 20, 1926.

61. "Shipments of Bananas from the Gatun Lake Area." *Canal Record*, January 28, 1925.

62. "First Steamer to Load Bananas in Gatun Lake," *Canal Record*, July 8, 1925.

63. "Banana Industry in Panama Has Good Future," *Star and Herald*, February 24, 1924.

64. "Agricultural Experiments," *Star and Herald*, January 23, 1926.

65. Marquardt, "Green Havoc," 49.

66. For example, Minor Keith purchased the corms that he initially used in Costa Rica from Carl Franc's plantations along the Chagres River. Companies around the canal during the 1920s also employed large-scale corm transfer practices.

67. "Virgin Banana Land [advertisement]," *Panama Times*, 1926 (no date).

68. W. H. Babbitt, banana farmer from the United States, to Meriwether L. Walker, governor of the Canal Zone, April 29, 1926, NACP RG 185, Entry 34, File 33.h.5, Bananas—Culture and Production in the Canal Zone and Vicinity.

69. "Banana Shipments from Canal Zone Ports," *Canal Record*, January 18, 1928.

70. "Banana Planters on Gatun Lake Face Crisis as Surplus Causes United Fruit to Decline Stems," *Panama American*, June 30, 1927.

71. N. A. Becker, land agent, to C. A. McIlvaine, executive secretary, December 30, 1922, NACP RG 185, Entry 34, File 33.b.51(4), Licensing Land in the CZ for Agricultural Purposes.

72. Memorandum for the engineer of maintenance, written by N. A. Becker, land agent, February 6, 1931, NACP RG 185, Entry 34, File 33.b.51(7).

73. Memorandum for the auditor, written by N. A. Becker, land agent, June 14, 1932, NACP RG 185, Entry 34, File 33.b.51(8).

74. Marquardt, "Green Havoc," 49.

75. "Banana Shipments from Canal Zone Ports," *Canal Record*, January 18, 1928.

76. March 26, 1932, NACP RG 185, Entry 34, File 33.b.51(8).

77. *Annual Report of the Governor of the Panama Canal for the Fiscal Year Ended June 30, 1932* (Washington, DC: Government Printing Office, 1932), 33.

Chapter 9

1. As historian Michael Conniff points out, estimates of mortality during the French canal project are unreliable, but this figure is widely cited; Conniff, *Black Labor on a White Canal: Panama, 1904–1981* (Pittsburgh, PA: University of Pittsburgh Press, 1985), 20.

2. *Diablo rojo* buses in Panama City were replaced by government buses in 2011.

3. During the period between the time I left the field in May 2009 and my return in July 2010, the construction of the Corredor Norte expressway between Panama City and Colón was completed. The toll road runs parallel to the Transístmica.

4. Government of Panama, 2000, Census. The national census of 2000 recorded 130 residents in Boquerón, compared to 183 residents in 1990.

Chapter 10

1. Harry Franck, *Zone Policeman 88: A Close Range Study of the Panama Canal and Its Workers* (New York: Hard Press, [1913] 2006), 69.

2. For a discussion of networked infrastructures, connection, and disconnection, see Stephen Graham and Simon Marvin, *Splintering Urbanism, Networked Infrastructure, Technological Mobilities, and the Urban Condition* (London: Routledge, 2001), 8–36. For a summary of how new transportation routes can obstruct mobility, see Jean-Paul Rodrigue, Claude Comtois, and Brian Slack, *The Geography of Transport Systems* (London: Routledge, 2013), 260.

3. "A Bridge over the Canal," *Panama Times*, September 26, 1925.

4. George Shiras, "Nature's Transformation at Panama: Remarkable Changes in Faunal and Physical Conditions in the Gatun Lake Region," *National Geographic* 28 (August 1915), 163.

5. Stephen Goddard, *Getting There: The Epic Struggle Between Road and Rail in the American Century* (Chicago: University of Chicago Press, 1994), 3.

6. Stanley Heckadon-Moreno, *Selvas Entre Dos Mares: Expediciones Científicas al Istmo de Panamá, Siglos XVIII-XX* (Panama City: Smithsonian Tropical Research Institute, 2005), 282–284.

7. "Wanted—Good Roads," *Star and Herald*, April 20, 1908.

8. Heckadon-Moreno, *Selvas Entre Dos Mares*, 282–284; George Roberts, *Investigación Económica de La República de Panamá* (Managua: Fundación UNO, [1932] 2006), 224; and Government of Panama (hereafter GOP), *Memoria Despacho de Fomento y Obras Publicas* (1920), ix–x.

9. On the discourse of releasing nature's energy through transportation infrastructure, see Richard White's discussion in the *Organic Machine*: "In Emersonian terms, putting land or water to work was opening, at least potentially, a new access to nature. Emerson had rejoiced in the 'magic' of railroad iron, 'its power to evoke the sleeping energies of land and water.'" Richard White, *The Organic Machine: The Remaking of the Columbia River* (New York: Hill and Wang, 1995), 35.

10. President Porras, "Mensaje dirigido por el Presidente de la Republica de Panamá a la Asamblea Nacional al inaugurar sus sesiones ordinarias el 1 de Septiembre de 1916" (Panama City: GOP); quoted in Ezer Vierba, *The Committee's Report: Punishment, Power and Subject in 20th Century Panamá*, (Dissertation, Department of History, Yale University, 2013), 36.

11. Panamanian historian Guillermo Castro Herrera argues that the 1936 treaty between the United States and Panama—known as the "Meat and Beer Treaty"—dramatically expanded cattle production and rural deforestation across the rural interior by opening, for the first time, the previously closed Canal Zone market to Panamanian agricultural and industrial products. The Remón-Eisenhower Treaty of 1955 further expanded demand by prohibiting Panamanian employees of the United States military and canal company from shopping in subsidized Canal Zone commissaries, forcing them to spend their money in Panamanian stores. See Guillermo Castro Herrera, "On Cattle and Ships: Culture, History and Sustainable Development in Panama," *Environment and History* 7, no. 2 (2001): 210–211.

12. "The Needs of Development," *The Workman*, April 18, 1925.

13. For a lively discussion of concrete and speed, see Michael Taussig, *My Cocaine Museum* (Chicago: University of Chicago Press, 2004), 159–172.

14. "Proposed Trans-Isthmian Highway," *Voice of the Panama Rotary Club* (1923).

15. Contract of 1867 between the United States of Colombia and the Panama Railroad Company, Article V.

16. Memorandum for the governor of the Canal Zone, written by N. A. Becker, land agent, December 2, 1931, National Archives at College Park, Maryland (hereafter NACP) RG 185, Entry 34, File 33.b.51(8), Licensing Land in CZ for Agricultural Purposes.

17. Memorandum for the governor of the Canal Zone, written by R. K. Morris, chief quartermaster, Sept. 9, 1921, NACP RG 185, Entry 34, File 33.b.51(2a)

18. "Copy of Report of Trans-Isthmian Highway Survey Delivered to Panama," *Canal Record*, July 26, 1933.

19. The shift in thinking about the canal's jungle defense was summarized in a 1935 canal defense report, which concluded: "The jungle is not by any means always an asset to the defense; on the contrary, by limiting fields of fire for all weapons, it frequently favors the attackers, who will presumably be in superior numbers. The fact is that the unbroken jungle can almost anywhere by traversed by considerable forces in any direction, with proper organization of working parties at a rate of 4 or 5 miles a day." See "Basic Project: The Project for the Defense of the Panama Canal," September 1, 1935, NACP RG 548, Box 16, Records of the US Army Forces in the Caribbean, 1935–1938, Defense Plans and Projects.

20. "Annexes to The Project for the Defense of the Panama Canal," NACP RG 584, Box 15, Defense Plans and Projects, 1925–1934, Records of US Army Forces in the Caribbean, Panama Canal Department.

21. "G-2 Appendix to the Basic War Plan, Panama Canal Department," p. 15–16, NACP RG 548, Box 16. Under war conditions, previously approved Panama Canal defense projects were to be constructed by the Army Corps of Engineers, civilian engineers, and the Canal Zone government. See "Engineer Annex to the Basic War Plan, Panama Canal Department," p. 6, NACP 548, Box 16.

22. Statement of Finley, US Department of State, May 31, 1940, NACP RG 59. Decimal, General Records of the Department of State.

23. NACP RG 59.Decimal, March 5, 1940.

24. John Carmody, Administrator of Federal Works Agency, to President Franklin Roosevelt, July 29, 1940, NACP RG 59.Decimal.

25. NACP RG 59.Decimal, June 17, 1940.

26. Hull to Smith, January 8, 1942, NACP RG 59.Decimal.

27. Draft Press Release, March 7, 1941, NACP RG 59.Decimal, NACP 59.Decimal.

28. "Engineer Annex to the Basic War Plan, Panama Canal Department," NACP RG 548.16, Records of US Army Forces in the Caribbean.

29. "Single Lane Done from Madden Dam to Randolph Road," *Panama American*, April 17, 1942.

30. US Federal Works to Department of State, April 28, 1941, NACP RG 59 .Decimal.

31. "Single Lane Done from Madden Dam to Randolph Road," *Panama American*, April 17, 1942.

32. Memo Chorrera-Rio Hato Highway, Public Roads and Department of State, August, 21, 1941, NACP RG 59.Decimal.

33. "Concrete Paving of Isthmian Highway Will Start Monday," *Panama American*, January 10, 1942.

34. "Trans-Isthmus Road to be Open to Civilians," *Panama American*, May 7, 1942.

35. "Single Lane Done from Madden Dam to Randolph Road," *Panama American*, April 17, 1942.

36. Ibid.

37. "President Inaugurates New Road Linking Colón with Capital and Rest of Republic," *Star and Herald*, April 16, 1943.

38. Rudolf Mrázek, *Engineers of Happy Land: Technology and Nationalism in a Colony* (Princeton, NJ: Princeton University Press, 2002), 10.

39. For information on settlement along the future highway route before construction, see GOP, *Memoria Despacho de Hacienda y Tesoro* (1936), 133. For detailed data on the construction of dirt roads from rural communities to the highway, see annual reports of road construction found in GOP, *Memoria Ministerio de Salubridad y Obras Publicas* (1942, 1943, and 1944).

40. Charles Bennett, *Human Influences on the Zoogeography of Panama* (Berkeley: University of California Press, 1968), 89.

Chapter 11

1. See Guillermo Castro Herrera, "Pro Mundi Beneficio: Elementos Para Una Historia Ambiental de Panamá," *Revista Tareas* 120 (May–August 2005): 82. Until the late twentieth century, logging was unregulated. Companies contracted workers to cut the most valuable trees around the canal. They floated the logs out by water for sale in the Canal Zone and elsewhere. See Stanley Heckadon-Moreno, *Los Sistemas de Producción Campesinos y los Recursos Naturales en la Cuenca del Canal* (Panama City: Ministerio de Desarrollo Agropecuario, Dirección Nacional de Recursos Naturales Renovables, y Agencia para el Desarrollo Internacional, 1981), 6. By 1915, natural

scientists in the Zone reported that nearly all of the forest within sight of the passing steamers and trains was secondary growth. See Henri Pittier, "Our Present Knowledge of the Forest Formations of the Isthmus of Panama," *Journal of Forestry* 16 (1918): 76–84.

2. Stanley Heckadon-Moreno, "Light and Shadows in the Management of the Panama Canal Watershed," in *The Rio Chagres: A Multidisciplinary Profile of a Tropical River Basin*, ed. Russell S. Harmon (New York: Kluwer Academic/Plenum Publishing, 2005), 37.

3. See Gina Porter, "Living in a Walking World: Rural Mobility and Social Equity Issues in Sub-Saharan Africa," *World Development* 30, no. 2 (2002): 285–300.

4. Stephen Gudeman, *The Demise of a Rural Economy: From Subsistence to Capitalism in a Latin American Village* (London: Routledge and Kegan Paul, 1978), 23. This point is also made in Stanley Heckadon-Moreno, *Panama's Expanding Cattle Front: The Santeno Campesinos and the Colonization of the Forests* (Dissertation, Department of Sociology, University of Essex, 1984), 141–144.

5. Alberto McKay, "Colonización de Tierras Nuevas En Panama," in *Agonía de la Naturaleza*, ed. Jaime Espinosa Gonzalez and Stanley Heckadon-Moreno (Panama City: IDIAP and Smithsonian Tropical Research Institute, 1985), 51–52.

6. Heckadon-Moreno, *Panama's Expanding Cattle Front*, 141.

7. Author personal communication with Francisco Herrera (Panamanian anthropologist), October 18, 2013.

8. Government of Panama (hereafter GOP), *Memoria Agricultura Comercio e Industrias* (1960), 283–284.

9. Heckadon-Moreno, *Panama's Expanding Cattle Front*, 166–167.

10. Alaka Wali, *Kilowatts and Crisis: Hydroelectric Power and Social Dislocation in Eastern Panama* (Boulder, CO: Westview Press, 1989), 12. On Che Guevara's description of the Alliance for Progress as the "latrinization" of Latin America, see Andre Gunder Frank and Martha Fuentes Frank, "The Development of Underdevelopment," in *Equity and Efficiency in Economic Development: Essays in Honour of Benjamin Higgins*, ed. Donald J. Savoie and Irving Brecher (Kingston, Ontario: McGill-Queen's University Press, 1992), 341–393.

11. This analysis draws on the critique of spontaneous colonization discourse laid out in Heckadon-Moreno, *Panama's Expanding Cattle Front*, 166–173.

12. Decentralized transportation has a long history in Panama; see chapter 5.

13. Wali, *Kilowatts and Crisis*, 12–13.

14. Heckadon-Moreno, *Panama's Expanding Cattle Front*, 147.

15. Candace Slater, "Amazonia as Edenic Narrative," in *Uncommon Ground: Rethinking the Human Place in Nature*, ed. William Cronon (New York: W.W. Norton and Co., 2005), 114–131.

16. For an insightful discussion of the historical distinction between "useful" and "non-useful" lands in Panama, see Guillermo Castro Herrera, "Panamá: Territorio, Sociedad y Desarrollo En La Perspectiva Del Siglo XXI," in *El Agua Entre Los Mares* (Panama City: Ciudad de Saber, 2007), 142–158.

17. Heckadon-Moreno, *Panama's Expanding Cattle Front*, 143.

18. Wali, 1989, *Kilowatts and Crisis*, 13.

19. Gloria Rudolf, *Panama's Poor: Victims, Agents, and Historymakers* (Gainesville: University Press of Florida, 1999), 137. I thank Ezer Vierba for bringing this quote to my attention. See Ezer Vierba, *The Committee's Report: Punishment, Power and Subject in 20th Century Panamá* (Dissertation, Department of History, Yale University, 2013), 424.

20. Heckadon-Moreno, *Panama's Expanding Cattle Front*, 148.

21. Stanley Heckadon-Moreno, "La Ganadería Extensiva y La Deforestación: Los Costos de Una Alternativa de Desarrollo," in *Agonía de La Naturaleza*, ed. Stanley Heckadon-Moreno and Jaime Espinos González (Panama City: Instituto de Investigación Agropecuaria de Panamá, 1985), 50–51.

22. Thomas E. Weil, Jan Knippers Black, Howard I. Blutstein, David S. McMorris, Frederick P. Munson, and Charles Townsend, *Area Handbook for Panama* (Washington, DC: Government Printing Office, 1972), 341–342.

23. GOP, *Memoria Obras Publicas* (1972).

24. GOP, *Memoria Agricultura Comercio e Industrias* (1967), 20B.

25. Vierba, *The Committee's Report*, 423.

26. Heckadon-Moreno, *Panama's Expanding Cattle Front*, 143.

27. On the problems that the intransigence of infrastructure—embedded with old ideologies and projects—poses during political shifts, see Stephen Collier, *Post-Soviet Social: Neoliberalism, Social Modernity, Biopolitics* (Princeton, NJ: Princeton University Press, 2012).

28. See Stephen Graham and Nigel Thrift, "Out of Order: Understanding Repair and Maintenance," *Theory, Culture & Society* 24, no. 3 (2007): 1–25. Notable exceptions include James Ferguson, *Expectations of Modernity: Myths and Meanings of*

Urban Life on the Zambian Copperbelt (Berkeley: University of California Press, 2006); Peter Redfield, *Space in the Tropics: From Convicts to Rockets in French Guiana* (Berkeley: University of California Press, 2000); and Ann Laura Stoler, "Imperial Debris: Reflections on Ruins and Ruination," *Cultural Anthropology* 23, no. 2 (2008): 191–219.

Chapter 12

1. Alfred W. Crosby, *Ecological Imperialism: The Biological Expansion of Europe, 900–1900* (Cambridge: Cambridge University Press, 1986), 7.

2. Although the title of "world's worst aquatic weed" is difficult to prove, water hyacinth is certainly a top contender due to the global scale of its invasion and significant hydroecological, economic, and social impacts. This description is used in Nuka Lata and Dubey Veenapani, "Response of Water Hyacinth Manure on Growth Attributes and Yield in Brassica Juncea," *Journal of Central European Agriculture* 12, no. 2 (2011): 336–343. On the ramifications of river control, including water hyacinth invasion, see W. M. Adams, *Green Development: Environment and Sustainability in the Third World* (London: Routledge, 1990), 223–228.

3. This summary draws on James A. Duke, *Handbook of Energy Crops* (unpublished, 1983); J. S. Hearne, "The Panama Canal's Aquatic Weed Problem," *Journal of Aquatic Plant Management* 5 (1966): 1–5; Dana R. Sanders, Sr., Russell F. Theriot, and Edwin A. Theriot, "Organisms Impacting Waterhyacinth in the Panama Canal," *Journal of Aquatic Plant Management* 20 (1982): 22–29. The claim the canal would be closed due to aquatic weeds within three to five years without control measures is widely cited, but rarely verified. In addition to Hearne's article, the interviews conducted with the Dredging Division's Aquatic Weed and Pollution Control Section that appear in Barbara Miller, "Airboats Defeat Old Panama Canal Foe," *Panama Canal Spillway*, September 3, 1982.

4. Secretary Samuel P. Langley, Smithsonian Institution, to President Theodore Roosevelt, March 21, 1904, Smithsonian Institution Archives (hereafter SIA) RU 45, Box 42, Folder 14, Office of the Secretary Records, 1903–1924.

5. Henri Pittier to Secretary Charles D. Walcott, Smithsonian Institution, January 1, 1911, SIA RU 45, Box 43, Folder 9.

6. A. S. Hitchcock to F. V. Coville, Smithsonian Institution, August 28, 1911, SIA RU 229, Box 11, Folder 1.

7. A. S. Hitchcock to F. V. Coville, September 10, 1911, SIA RU 229, Box 11, Folder 1.

8. See National Archives at College Park, Maryland (hereafter NACP) RG 185, Entry 30, File 33.h.4(1), Aquatic Plants (Hyacinth, Chara, Naias, Cabomba, etc.) in Canal Waters; Studies, Reports, Methods of Control and Elimination to Prevent Mosquito Breeding and Obstructions to Navigation, etc.

9. These efforts are documented in the correspondence found in NACP 185, Entry 30, File 33.h.4(1), Aquatic Plants in Canal Waters.

10. Otto Lutz, professor of natural science at the National Institute of Panama, to George Goethals, chairman and chief engineer, May 16, 1913, NACP 185, Entry 30 File, 33.h.4(1)

11. Inspector to George Goethals, May 11, 1913, NACP 185, Entry 30 File, 33.h.4(1).

12. Resident engineer to George Goethals, Dec. 18, 1913, NACP 185, Entry 30 File, 33.h.4(1).

13. Lewis B. Bates, acting chief of the Board of Health Laboratory, to acting chief sanitary officer, Dec. 24, 1913, NACP 185, Entry 30, File 33.h.4(1).

14. Law proposed by Mr. Comber, resident engineer, to Chester Harding, acting governor of the Canal Zone, May 17, 1915, NACP 185, Entry 34, File 33.h 4(2), Aquatic Plants in Canal Waters; and response by Frank Feuille, chief council, to Chester Harding, May 20, 1915, NACP 185, Entry 34, File 33.h 4(2).

15. Reverend Thorbourne, founder of the West Indian community of New Providence, to Chester Harding, governor of the Canal Zone, June 21, 1915, NACP 185, Entry 30 File, 33.h.4(1), Aquatic Plants in Canal Waters.

16. O. E. Malsbury, assistant engineer, to Mr. Douglas, engineer of maintenance, March 7, 1919, NACP 185, Entry 30 File, 33.h.4(2).

17. Mr. Douglas, engineer of maintenance, to Chester Harding, governor of the Canal Zone, March 13, 1919, NACP 185, Entry 30 File, 33.h.4(2).

18. "Adventitious," *American Heritage College Dictionary* (2002).

Chapter 13

1. This section summarizes journalist Alan Weisman's discussion of the disassembly of the Panama Canal, part of a larger project on how—and how quickly—various artifacts of the human built world that we take for granted could fall apart; Alan Weisman, *The World Without Us* (New York: Picador, 2007).

2. Clifford Geertz, "The Wet and the Dry: Traditional Irrigation in Bali and Morocco," *Human Ecology* 1, no. 1 (1972): 38.

3. Cultures of circulation are theorized in terms of "demanding environments" in Dilip P. Gaonkar and Elizabeth A. Povinelli, "Technologies of Public Forms: Circulation, Transfiguration, Recognition," *Public Culture* 15, no. 3 (2003): 395.

4. Mike Davis, *Ecology of Fear: Los Angeles and the Imagination of Disaster* (New York: Vintage, 1999), 8.

5. Gary L. Strang, "Infrastructure as Landscape," *Places* 10, no. 3 (1996): 11.

6. Pierre Bélanger, "Landscape as Infrastructure," *Landscape Journal* 28, no. 1 (2009): 79–95.

Bibliography

Abbot, Henry. *Problems of the Panama Canal.* New York: The Macmillan Company, 1905.

Adams, Frederick Upham. *Conquest of the Tropics: The Story of the Creative Enterprises Conducted by the United Fruit Company.* Garden City, NY: Doubleday, Page and Co., 1914.

Adams, W. M. *Green Development: Environment and Sustainability in the Third World.* London: Routledge, 1990.

Akrich, Madeleine. "The De-Scription of Technical Objects." In *Shaping Technology/ Building Society: Studies in Sociotechnical Change,* edited by Wiebe E. Bijker and John Law, 205–224. Cambridge, MA: MIT Press, 1992.

Allen, Paul. "The Timber Woods of Panama." *Ceiba* 10, no. 1 (1964): 17–61.

Anand, Nikhil. "Pressure: The Politechnics of Water Supply in Mumbai." *Cultural Anthropology* 26, no. 4 (2011): 542–564.

Andreassian, Vazken. "Waters and Forests: From Historical Controversy to Scientific Debate." *Journal of Hydrology* 291, no. 1 (2004): 1–27.

Anguizola, Gustave. "Negroes in the Building of the Panama Canal." *Phylon* 29, no. 4 (1968): 351–359.

Appadurai, Arjun. "Disjuncture and Difference in the Global Cultural Economy." In *Modernity at Large: Cultural Dimensions of Globalization,* 27–47. Minneapolis: University of Minnesota Press, [1990] 1996.

Bakenhus, Reuben E., Harry S. Knapp, and Emory R. Johnson. *The Panama Canal: Comprising Its History and Construction, and Its Relation to the Navy, International Law and Commerce.* New York: John Wiley and Sons, 1915.

Bakker, Karen. "Water: Political, Biopolitical, Material." *Social Studies of Science* 42, no. 4 (2012): 616–623.

Barry, Andrew. "Technological Zones." *European Journal of Social Theory* 9, no. 2 (2006): 239–253.

Bates, C. G., and A. J. Henry. "Forest and Streamflow Experiment at Wagon Wheel Gap, Colorado." *Monthly Weather Review Supplement* 3. Washington, DC: USDA Weather Bureau, 1928.

Bateson, Gregory. *Steps to an Ecology of Mind.* Chicago: University of Chicago Press, [1972] 2000.

Batt, William H. "Infrastructure: Etymology and Import." *Journal of Professional Issues in Engineering* 110, no. 1 (1984): 1–6.

Bélanger, Pierre. "Landscape as Infrastructure." *Landscape Journal* 28, no. 1 (2009): 79–95.

Bennett, Charles. *Human Influences on the Zoogeography of Panama.* Berkeley: University of California Press, 1968.

Bennett, Hugh H., and William A. Taylor. *The Agricultural Possibilities of the Canal Zone.* Washington, DC: Government Printing Office, 1912.

Bennett, Jane. *Vibrant Matter: A Political Ecology of Things.* Durham, NC: Duke University Press, 2010.

Bennett, Tony, and Patrick Joyce, eds. *Cultural Studies, History and the Material Turn.* New York: Routledge, 2010.

Blaikie, Piers, and Harold Brookfield, eds. *Land Degradation and Society.* London: Metheun, 1987.

Blas Tejeira, Gil. *Pueblos Perdidos.* Panama City: Editorial Universitaria, [1962] 2003.

Bosch, J. M., and J. D. Hewlett. "A Review of Catchment Experiments to Determine the Effect of Vegetation Changes on Water Yield and Evapotranspiration." *Journal of Hydrology* 55, no. 1 (1982): 3–23.

Bourgois, Philippe. *Ethnicity at Work: Divided Labor on a Central American Banana Plantation.* Baltimore, MD: The Johns Hopkins University Press, 1989.

Bourgois, Philippe. "100 Years of United Fruit Company Letters." In *Banana Wars: Power, Production, and History in the Americas,* edited by Steve Strifler and Mark Moberg, 103–144. Durham, NC: Duke University Press, 2003.

Bowker, Geoffrey C. *Science on the Run: Information Management and Industrial Geophysics at Schlumberger, 1920–1940.* Cambridge, MA: MIT Press, 1994.

Bowker, Geoffrey C., and Susan Leigh Star. *Sorting Things Out: Classification and Its Consequences.* Cambridge, MA: MIT Press, 1999.

Bruijnzeel, Leendert A. "Hydrological Functions of Tropical Forests: Not Seeing the Soil for the Trees?" *Agriculture, Ecosystems and Environment* 104, no. 1 (2004): 185–228.

Budowski, Gerardo. "Forestry Training in Latin America." *Caribbean Forester* (January-June 1961): 33–38.

Busch, Lawrence. *Standards: Recipes for Reality*. Cambridge, MA: MIT Press, 2011.

Calder, Ian. "Forests and Water—What We Know and What We Need to Know." *Science for Nature Symposium*. Washington, DC, 2006.

Callon, Michel. "Some Elements of a Sociology of Translation: Domestication of the Scallops and the Fishermen of St. Brieuc Bay." In *Power, Action and Belief: A New Sociology of Knowledge?*, edited by John Law, 196–223. London: Routledge, 1986.

Callon, Michel. "Techno-Economic Networks and Irreversibility." In *A Sociology of Monsters: Essays on Power, Technology and Domination*, edited by John Law, 132–161. London: Routledge, 1995.

Calvo Sucre, Alberto E. *Análisis Histórico del Desarrollo de la Salud Publica en la Republica de Panamá, 1903–2006*. Panama City: Editorial Universitaria Carlos Manuel Gasteazoro, 2007.

Carroll, Patrick. *Science, Culture, and Modern State Formation*. Berkeley: University of California Press, 2006.

Castells, Manuel. *The Rise of the Network Society*. Hoboken, NJ: Wiley, 1996.

Castillero Calvo, Alfredo. "Transistmo y Dependencia: El Caso del Istmo de Panamá." *Loteria* 211 (1973): 25–56.

Castillero Calvo, Alfredo. *La Ruta Transistmica y las Comunicaciones Marítimas Hispanas Siglos XVI a XIX*. Panama City: Ediciones Nari, 1984.

Castro Herrera, Guillermo. "On Cattle and Ships: Culture, History and Sustainable Development in Panama." *Environment and History* 7, no. 2 (2001): 201–217.

Castro Herrera, Guillermo. "Panamá: Agua y Desarrollo en Vísperas del Segundo Siglo." *Revista Tareas* 114 (May–August 2003): 21–52.

Castro Herrera, Guillermo. "Pro Mundi Beneficio: Elementos para una Historia Ambiental de Panamá." *Revista Tareas* 120 (May–August 2005): 81–112.

Castro Herrera, Guillermo. *El Agua Entre los Mares*. Panama City: Ciudad de Saber, 2007.

Chomsky, Aviva. *West Indian Workers and the United Fruit Company in Costa Rica, 1870–1940*. Baton Rouge: Louisiana State University Press, 1996.

Clegg, Peter. *The Caribbean Banana Trade: From Colonialism to Globalization*. New York: Palgrave Macmillan, 2002.

Coates, Anthony G. "The Forging of Central America." In *Central America: A Natural and Cultural History*, edited by Anthony G. Coates, 1–37. New Haven, CT: Yale University Press, 1997.

Coates, Anthony G., Jeremy B. C. Jackson, Laurel S. Collins, Thomas M. Cronin, Harry J. Dowsett, Laurel M. Bybell, Peter Jung, and Jorge A. Obando. "Closure of the Isthmus of Panama: The Near-Shore Marine Record of Costa Rica and Western Panama." *Geological Society of American Bulletin* 104, no. 7 (1992): 814–828.

Collier, Stephen. *Post-Soviet Social: Neoliberalism, Social Modernity, Biopolitics*. Princeton, NJ: Princeton University Press, 2012.

Collier, Stephen J., and Aihwa Ong. "Global Assemblages, Anthropological Problems." In *Global Assemblages: Technology, Politics, and Ethics as Anthropological Problems*, edited by Aihwa Ong and Stephen J. Collier, 3–21. Malden, MA: Blackwell, 2005.

Collins, Laurel S., Anthony G. Coates, William A. Berggren, Marie Pierre Aubry, and Jihun Zhang. "The Late Miocene Panama Isthmian Strait." *Geology* 24, no. 8 (2006): 687–690.

Conniff, Michael L. *Black Labor on a White Canal: Panama, 1904–1981*. Pittsburgh, PA: University of Pittsburgh Press, 1985.

Cooke, Richard. "The Native Peoples of Central America during Precolumbian and Colonial Times." In *Central America: A Natural and Cultural History*, edited by Anthony G. Coates, 137–176. New Haven, CT: Yale University Press, 1999.

Cooke, Richard, Dolores Piperno, Anthony J. Ranere, Karen Clary, Patricia Hansell, Storrs Olson, Valerio L. Wilson, and Doris Weiland. "La Influencia de las Poblaciones Humanas Sobre los Ambientes Terrestres de Panama Entre el 10,000 A.C. y el 500 D.C." In *Agonia de la Naturaleza*, edited by Stanley Heckadon-Moreno and Jaime Espinosa Gonzalez, 3–25. Panama City: IDIAP and STRI, 1985.

Cortéz, Rosa María. "Características Generales de la Población." In *La Cuenca del Canal de Panama: Actas de los Seminarios-Talleres*, edited by Stanley Heckadon-Moreno, 45–52. n.p.: Government of Panama, 1986.

Cronon, William. *Nature's Metropolis: Chicago and the Great West*. New York: W. W. Norton and Co., 1991.

Cronon, William. "Kennecott Journey: The Paths Out of Town." In *Under an Open Sky: Rethinking America's Western Past*, edited by William Cronon, George Miles, and Jay Gitlin, 28–51. New York: W. W. Norton and Co., 1992.

Crosby, Alfred. *Ecological Imperialism: The Biological Expansion of Europe, 900–1900.* Cambridge: Cambridge University Press, 1986.

Cummings, Laurence J. *Forestry in Panama.* Panama City: SICAP, 1956.

Davis, Mike. *Ecology of Fear: Los Angeles and the Imagination of Disaster.* New York: Vintage, 1999.

Denevan, William. "The Pristine Myth: The Landscape of the Americas in 1492." *Annals of the Association of American Geographers* 82, no. 3 (1992): 369–385.

Dodds, Gordon B. "The Stream-Flow Controversy: A Conservation Turning Point." *Journal of American History* 56, no. 1 (1969): 59–69.

Edwards, Paul. "Infrastructure and Modernity: Force, Time, and Social Organization in the History of Sociotechnical Systems." In *Modernity and Technology*, edited by Thomas, J. Misa, Philip Brey, and Andrew Feenberg, 185–226. Cambridge, MA: MIT Press, 2003.

Edwards, Paul. *A Vast Machine: Computer Models, Climate Data, and the Politics of Global Warming.* Cambridge, MA: MIT Press, 2010.

Edwards, Paul, Steven J. Jackson, Geoffrey C. Bowker, and Cory P. Knobel. *Understanding Infrastructure: Dynamics, Tensions, and Design.* Report of a Workshop on History and Theory of Infrastructure: Lessons for New Scientific Cyberinfrastructures. Ann Arbor, University of Michigan, 2007.

Egyedi, Tineke. "Infrastructure Flexibility Created by Standardized Gateways: The Cases of XML and the ISO Container." *Knowledge, Technology & Policy* 14, no. 3 (2001): 41–54.

Escobar, Arturo. *Encountering Development: The Making and Unmaking of the Third World.* Princeton, NJ: Princeton University Press, 1995.

Ferguson, James. *Global Shadows: Africa in the Neoliberal World Order.* Durham, NC: Duke University Press, 1999.

Ferguson, James. *Expectations of Modernity: Myths and Meanings of Urban Life on the Zambian Copperbelt.* Berkeley: University of California Press, 2006.

Ferguson, James, and Akhil Gupta. "Spatializing States: Toward an Ethnography of Neoliberal Governmentality." *American Ethnologist* 29, no. 4 (2002): 981–1002.

Fernández de Oviedo, Gonzalo. *Natural History of the West Indies.* Chapel Hill, NC: The Orange Printshop, [1526] 1959.

Fischer, Michael. "Technoscientific Infrastructures and Emergent Forms of Life: A Commentary." *American Anthropologist* 107, no. 1 (2005): 55–61.

Franck, Harry. *Zone Policeman 88: A Close Range Study of the Panama Canal and Its Workers*. New York: Hard Press, [1913] 2006.

Frank, Andre Gunder. *Capitalism and Underdevelopment in Latin America*. New York: Monthly Review Press, 1969.

Frank, Andre Gunder, and Martha Fuentes Frank. "The Development of Underdevelopment." In *Equity and Efficiency in Economic Development: Essays in Honour of Benjamin Higgins*, edited by Donald J. Savoie and Irving Brecher, 341–393. Kingston, Ontario: McGill-Queen's University Press, 1992.

Frenkel, Stephen. "Geographical Representations of the 'Other': The Landscape of the Panama Canal Zone." *Journal of Historical Geography* 28, no. 1 (2002): 85–99.

Gandy, Matthew. *Concrete and Clay: Reworking Nature in New York City*. Cambridge, MA: MIT Press, 2002.

Gaonkar, Dilip P., and Elizabeth A. Povinelli. "Technologies of Public Forms: Circulation, Transfiguration, Recognition." *Public Culture* 15, no. 3 (2003): 385–397.

Gause, Frank, and Charles C. Carr. *The Story of Panama: The New Route to India*. Boston, MA: Silver, Burdett and Co., 1912.

Geertz, Clifford. "The Wet and the Dry: Traditional Irrigation in Bali and Morocco." *Human Ecology* 1, no. 1 (1972): 23–39.

Gillem, Mark L. *America Town: Building the Outposts of Empire*. Minneapolis: University of Minnesota Press, 2007.

Goddard, Stephen B. *Getting There: The Epic Struggle Between Road and Rail in the American Century*. Chicago: University of Chicago Press, 1994.

Goethals, George. *Government of the Canal Zone*. Princeton, NJ: Princeton University Press, 1915.

Gorgas, William. *Sanitation in Panama*. New York: D. Appleton and Co., 1915.

Graham, Stephen and Simon Marvin. *Splintering Urbanism: Networked Infrastructure, Technological Mobilities and the Urban Condition*. London: Routledge, 2001.

Graham, Stephen, and Nigel Thrift. "Out of Order: Understanding Repair and Maintenance." *Theory, Culture & Society* 24, no. 3 (2007): 1–25.

Greene, Julie. *The Canal Builders*. New York: Penguin, 2009.

Grossman, Lawrence. "The Political Ecology of Banana Exports and Local Food Production in St. Vincent, Eastern Caribbean." *Annals of the Association of American Geographers* 83, no. 2 (1993): 347–367.

Gudeman, Stephen. *The Demise of a Rural Economy: From Subsistence to Capitalism in a Latin American Village*. London: Routledge, [1978] 1988.

Hamilton, Lawrence S., and Peter N. King. *Tropical Forested Watersheds: Hydrologic and Soils Response to Major Uses or Conversions*. Boulder, CO: Westview Press, 1983.

Hannerz, Ulf. "Notes on the Global Ecumene." *Public Culture* 1, no. 2 (1989): 66–75.

Harvey, Penelope. "The Materiality of State Effects: An Ethnography of a Road in the Peruvian Andes." In *State Formation: Anthropological Perspectives*, edited by Christian Krohn-Hansen, Knut G. Nustad, 216–247. London: Pluto Press, 2005.

Hearne, J. S. "The Panama Canal's Aquatic Weed Problem." *Journal of Aquatic Plant Management* 5 (1966): 1–5.

Heckadon-Moreno, Stanley. *Los Sistemas de Producción Campesinos y los Recursos Naturales en la Cuenca del Canal*. n.p.: Government of Panama, 1981.

Heckadon-Moreno, Stanley. *Panama's Expanding Cattle Front: The Santeno Campesinos and the Colonization of the Forests*. Dissertation, Department of Sociology, University of Essex, 1984.

Heckadon-Moreno, Stanley. "La Ganadería Extensiva y la Deforestación: Los Costos de una Alternativa de Desarrollo." In *Agonía de la Naturaleza*, edited by Stanley Heckadon-Moreno and Jaime Espinos González, 45–62. Panama City: Instituto de Investigación Agropecuaria de Panamá, 1985.

Heckadon-Moreno, Stanley, ed. *La Cuenca del Canal de Panama: Actas de los Seminarios-Talleres*. n.p.: Government of Panama, 1986.

Heckadon-Moreno, Stanley. "Impact of Development on the Panama Canal Environment." *Journal of Interamerican Studies and World Affairs* 35, no. 3 (1993): 129–148.

Heckadon-Moreno, Stanley. "Light and Shadows in the Management of the Panama Canal Watershed." In *The Rio Chagres: A Multidisciplinary Profile of a Tropical River Basin*, edited by Russell S. Harmon, 28–44. New York: Kluwer Academic/Plenum Publishing, 2005.

Heckadon-Moreno, Stanley. *Selvas Entre Dos Mares: Expediciones Científicas al Istmo de Panamá, Siglos XVIII–XX*. Panama City: Smithsonian Tropical Research Institute, 2005.

Heckadon-Moreno, Stanley. *Cuando se Acaban los Montes: Los Campesinos Santeños y la Colonización de Tonosi*. Panama City: Editorial Universitaria Carlos Manuel Gasteazoro, 2006.

Holdridge, L. R., and Gerardo Budowski. "Report on an Ecological Survey of the Republic of Panama." *Caribbean Forester* 17 (1956): 92–110.

Hooper, Bruce. *Integrated River Basin Governance: Learning from International Experience*. London: IWA Publishing, 2005.

Hricko, Andrea. "Progress and Pollution: Port Cities Prepare for the Panama Canal Expansion." *Environmental Health Perspectives* 120, no. 12 (2012): A471.

Jaén Suárez, Omar. *Análisis Regional y Canal de Panamá: Ensayos Geográficos*. Panama City: Editorial Universitaria, 1981.

Jaén Suárez, Omar. *Hombres y Ecología En Panamá*. Panama City: Editorial Universitaria, Smithsonian Tropical Research Institute, 1981.

Jenkins, Virginia Scott. *Bananas: An American History*. Washington, DC: Smithsonian Institution, 2000.

Joseph, Gilbert M., Catherine C. LeGrand, and Ricardo D. Salvatore, eds. *Close Encounters of Empire: Writing the Cultural History of US-Latin American Relations*. Durham, NC: Duke University Press, 1998.

Kaimowitz, David. "Useful Myths and Intractable Truths: The Politics of the Links between Forests and Water in Central America." In *Forests, Water and People in the Humid Tropics: Past, Present and Future Hydrological Research for Integrated Land and Water Management*, edited by Michael Bonnell and Leendert A. Bruijnzeel, 86–98. Cambridge: Cambridge University Press, 2004.

Kepner, Charles David, Jr., and Jay Henry Soothill. *The Banana Empire: A Case Study of Economic Imperialism*. New York: The Vanguard Press, 1935.

Kirkpatrick, R. Z. "Madden Dam Will Insure Water Supply for Gatun Lake Development." *Iowa Engineer* (March 1934): 84–85.

Kirsch, Scott, and Don Mitchell. "Earth-Moving as the 'Measure of Man': Edward Teller, Geographical Engineering, and the Matter of Progress." *Social Text* 54, no. 16 (1998): 100–134.

Kittredge, Joseph. *Forest Influences*. New York: McGraw-Hill, 1948.

Koeppel, Dan. *Banana: The Fate of the Fruit That Changed the World*. New York: Hudson Street Press, 2008.

LaFeber, Walter. *The Panama Canal: The Crisis in Historical Perspective*. New York: Oxford University Press, 1978.

Langdon, Robert. "The Banana as a Key to Early American and Polynesian History." *Journal of Pacific History* 28, no. 1 (1993): 15–35.

Larkin, Brian. "The Politics and Poetics of Infrastructure." *Annual Review of Anthropology* 42, no. 1 (2013): 327–343.

Larson, Curtis. "Erosion and Sediment Yields as Affected by Land Use and Slope in the Panama Canal Watershed." In *Third World Congress on Water Resources*, vol. 3–4, 1086–1095. Mexico City: International Water Resources Association, 1979.

Lata, Nuka, and Dubey Veenapani, "Response of Water Hyacinth Manure on Growth Attributes and Yield in Brassica Juncea." *Journal of Central European Agriculture* 12, no. 2 (2011): 336–343.

Latour, Bruno. *Pandora's Hope: Essays on the Reality of Science Studies*. Cambridge, MA: Harvard University Press, 1999.

Latour, Bruno. *Reassembling the Social: An Introduction to Actor-Network-Theory*. Oxford: Oxford University Press, 2005.

Law, John. "Objects and Spaces." *Theory, Culture and Society* 19, no. 5/6 (2002): 91–105.

Law, John, and Annemarie Mol. "Situating Technoscience: An Inquiry into Spatialities." *Environment and Planning D: Society and Space* 19, no. 5 (2001): 609–621.

Lefebvre, Henri. *The Production of Space*. Oxford: Blackwell, [1974] 1991.

Levinson, Marc. *The Box: How the Shipping Container Made the World Smaller and the World Economy Bigger*. Princeton, NJ: Princeton University Press, 2006.

Lewis, Lancelot S. *The West Indian in Panama: Black Labor in Panama, 1850–1914*. Washington, DC: University Press of America, 1980.

Lindsay-Poland, John. *Emperors in the Jungle: The Hidden History of the US in Panama*. Durham, NC: Duke University Press, 2003.

Lowenthal, David. *George Perkins Marsh: Prophet of Conservation*. Seattle: University of Washington Press, 2000.

Mack, Gerstle. *The Land Divided: A History of the Panama Canal and Other Isthmian Canal Projects*. New York: Knopf, 1944.

Major, John. *Prize Possession: The United States and the Panama Canal, 1903–1979*. Cambridge: Cambridge University Press, 1993.

Mann, Charles C. *1491: New Revelations of the Americas before Columbus*. New York: Vintage Books, 2006.

Marquardt, Steve. "'Green Havoc': Panama Disease, Environmental Change, and Labor Process in the Central American Banana Industry." *American Historical Review* 106, no. 1 (2001): 49–80.

Marsh, George Perkins. *Man and Nature*. Seattle: University of Washington Press, [1864] 2003.

Masco, Joseph. *The Nuclear Borderlands: The Manhattan Project in Post–Cold War New Mexico*. Princeton, NJ: Princeton University Press, 2006.

Mastellari Navarro, Jorge. *Zona del Canal: Analogía de una Colonia*. Panama City: n.p., 2003.

McCulloch, James S. G., and Mark Robinson. "History of Forest Hydrology." *Journal of Hydrology* 150, no. 2 (1993): 189–216.

McCullough, David. *The Path Between the Seas: The Creation of the Panama Canal, 1870–1914.* New York: Simon and Schuster, [1977] 2001.

McGuinness, Aims. *Path of Empire: Panama and the California Gold Rush.* Ithaca, NY: Cornell University Press, 2008.

McKay, Alberto. "Colonización de Tierras Nuevas En Panama." In *Agonía de la Naturaleza*, edited by Jaime Espinosa Gonzalez and Stanley Heckadon-Moreno, 45–62. Panama City: IDIAP and Smithsonian Tropical Research Institute, 1985.

Mitchell, Timothy. *Rule of Experts: Egypt, Techno-Politics, Modernity.* Berkeley: University of California Press, 2002.

Mitchell, Timothy. *Carbon Democracy: Political Power in the Age of Oil.* London: Verso, 2013.

Molle, Francois. "River-Basin Planning and Management: The Social Life of a Concept." *Geoforum* 40, no. 3 (2009): 484–494.

Moore, Donald S. *Suffering for Territory: Race, Place, and Power in Zimbabwe.* Durham, NC: Duke University Press, 2005.

Mrázek, Rudolf. *Engineers of Happy Land: Technology and Nationalism in a Colony.* Princeton, NJ: Princeton University Press, 2002.

Mukerji, Chandra. *Impossible Engineering: Technology and Territoriality on the Canal Du Midi.* Princeton, NJ: Princeton University Press, 2009.

Müller-Wille, Staffen. "Introduction." In *Musa Cliffortiana: Clifford's Banana Plant*, by Carl Linnaeus, 15–67. Ruggell, Liechtenstein: A.R.G. Ganternet Verlag K.G., 2007.

Otis, F. N. *Illustrated History of the Panama Railroad.* Pasadena, CA: Socio-Technical Books, [1861] 1971.

Parker, Matthew. *Panama Fever: The Battle to Build the Panama Canal.* London: Hutchinson, 2007.

Patiño Mejía, Alberto. "La Corporación del Valle del Cauca Promotora de Desarrollo Rural." In *La Cuenca del Canal de Panama: Actas de los Seminarios-Talleres*, edited by Stanley Heckadon-Moreno, 259–272. n.p.: Government of Panama, 1986.

Peet, Richard, and Elaine Hartwick. *Theories of Development: Contentions, Arguments, Alternatives.* New York: Guilford Press, 2009.

Pereira Jiménez, Bonifacio. *Biografía del Río Chagres.* Panama City: Imprenta Nacional, 1964.

Pinzon, Luis, and Jose Esturain. "Vigilancia de los Bosques." In *La Cuenca del Canal de Panama: Actas de los Seminarios-Talleres*, edited by Stanley Heckadon-Moreno, 205–215. n.p.: Government of Panama, 1986.

Pisani, Donald J. "A Conservation Myth: The Troubled Childhood of the Multiple-Use Idea." *Agricultural History* 76, no. 2 (2000): 154–171.

Pittier, Henri. "Our Present Knowledge of the Forest Formations of the Isthmus of Panama." *Journal of Forestry* 16, no. 1 (1918): 76–84.

Portelli, Alessandro. *The Death of Luigi Trastulli and Other Stories: Form and Meaning in Oral History*. Albany: State University of New York Press, 1991.

Porter, Gina. "Living in a Walking World: Rural Mobility and Social Equity Issues in Sub-Saharan Africa." *World Development* 30, no. 2 (2002): 285–300.

Powell, John Wesley. "Institutions for the Arid Lands." *Century Magazine* 40, no. 1 (May 1890): 111–116.

Prebisch, Raul. *International Economics and Development*. New York: Academic Press, 1972.

Pritchard, Sara B. *Confluence: The Nature of Technology and the Remaking of the Rhone*. Cambridge, MA: Harvard University Press, 2011.

Raffles, Hugh. "The Uses of Butterflies." *American Ethnologist* 28, no. 3 (2001): 513–548.

Redfield, Peter. "Beneath a Modern Sky: Space Technology and Its Place on the Ground." *Science, Technology, & Human Values* 21, no. 3 (1996): 251–274.

Redfield, Peter. *Space in the Tropics: From Convicts to Rockets in French Guiana*. Berkeley: University of California Press, 2000.

Reuss, Martin, and Stephen H. Cutcliffe. *The Illusory Boundary: Environment and Technology in History*. Charlottesville: University of Virginia Press, 2010.

Richardson, Bonham C. *Panama Money in Barbados, 1900–1920*. Knoxville: University of Tennessee Press, 1985.

Roberts, George E. *Investigación Económica de la República de Panamá*. Managua: Fundación UNO, [1932] 2006.

Robinson, Frank H. *A Report on the Panama Canal Rain Forests*. n.p.: Panama Canal Commission, 1985.

Roche, Julian. *The International Banana Trade*. Boca Raton, FL: CRC Press, 1998.

Rockefeller, Stuart Alexander. "Flow." *American Anthropologist* 52, no. 4 (2011): 557–578.

Rodrigue, Jean-Paul. "Straits, Passages and Chokepoints: A Maritime Geostrategy of Petroleum Distribution." *Cahiers de Géographie du Québec* 48 (December 2004): 357–374.

Rodrigue, Jean-Paul, Claude Comtois, and Brian Slack. *The Geography of Transport Systems*. London: Routledge, 2013.

Rostow, Walter W. *The Stages of Economic Growth: A Non-Communist Manifesto*. Cambridge: Cambridge University Press, 1960.

Rowe, Leo S. "The Work of the Joint International Commission on Panamanian Land Claims." *The American Journal of International Law* 8, no. 4 (1914): 738–757.

Rudolf, Gloria. *Panama's Poor: Victims, Agents, and Historymakers*. Gainesville: University Press of Florida, 1999.

Saberwal, Vasant K. "Science and the Desiccationist Discourse of the 20th Century." *Environment and History* 4, no. 3 (1998): 309–343.

Sack, Robert David. *Human Territoriality: Its Theory and History*. Cambridge: Cambridge University Press, 1986.

Sanders, Dana R., Sr., Russell F. Theriot, and Edwin A. Theriot. "Organisms Impacting Waterhyacinth in the Panama Canal." *Journal of Aquatic Plant Management* 20 (1982): 22–29.

Scott, James C. *Seeing Like a State: How Certain Schemes to Improve the Human Condition Have Failed*. New Haven, CT: Yale University Press, 1998.

Scoullar, William T. *Libro Azul de Panama*. Panama City: Imprenta Nacional, 1917.

Shallat, Todd. *Structures in the Stream: Water, Science, and the Rise of the US Army Corps of Engineers*. Austin: University of Texas Press, 1994.

Shiras, George. "Nature's Transformation at Panama: Remarkable Changes in Faunal and Physical Conditions in the Gatun Lake Region." *National Geographic* 28 (August 1915): 159–194.

Slater, Candace. "Amazonia as Edenic Narrative." In *Uncommon Ground: Rethinking the Human Place in Nature*, edited by William Cronon, 114–131. New York: W. W. Norton and Co., 2005.

Slocum, Karla. *Free Trade and Freedom: Neoliberalism, Place, and Nation in the Caribbean*. Ann Arbor: University of Michigan Press, 2006.

Smith, T. C. "The Drainage Basin as an Historical Unit for Human Activity." In *Introduction to Geographical Hydrology*, edited by R. J. Chorley, 20–29. London: Methuen, 1971.

Soluri, John. *Banana Cultures: Agriculture, Consumption, and Environmental Change in Honduras and the United States*. Austin: University of Texas Press, 2005.

Star, Susan Leigh. "The Ethnography of Infrastructure." *American Behavioral Scientist* 43, no. 3 (1999): 377–391.

Star, Susan Leigh, and James R. Griesemer. "Institutional Ecology, 'Translations' and Boundary Objects: Amateurs and Professionals in Berkeley's Museum of Vertebrate Zoology, 1907–39." *Social Studies of Science* 19, no. 3 (1989): 387–420.

Star, Susan Leigh and Karen Ruhleder. "Steps Toward an Ecology of Infrastructure: Design and Access for Large Information Spaces." *Information Systems Research* 7, no. 1 (1996): 111–134.

Stoler, Ann Laura. "Tense and Tender Ties: The Politics of Comparison in North American History and (Post) Colonial Studies." *Journal of American History* 88, no. 3 (2001): 829–865.

Stoler, Ann Laura. "Imperial Debris: Reflections on Ruins and Ruination." *Cultural Anthropology* 23, no. 2 (2008): 191–219.

Strang, Gary L. "Infrastructure as Landscape." *Places* 10, no. 3 (1996): 8–15.

Striffler, Steve, and Mark Moberg, eds. *Banana Wars: Power, Production, and History in the Americas*. Durham, NC: Duke University Press, 2003.

Sutter, Paul S. "Nature's Agents or Agents of Empire? Entomological Workers and Environmental Change During the Construction of the Panama Canal." *Isis* 98, no. 4 (2007): 724–754.

Sutter, Paul S. "Tropical Conquest and the Rise of the Environmental Management State: The Case of US Sanitary Efforts in Panama." In *Colonial Crucible: Empire in the Making of the Modern American State*, edited by Alfred W. McCoy and Francisco A. Scarano, 317–326. Madison: University of Wisconsin Press, 2009.

Swyngedouw, Erik. *Social Power and the Urbanization of Water: Flows of Power*. New York: Oxford University Press, 2004.

Szok, Peter A. *"La Última Gaviota": Liberalism and Nostalgia in Early Twentieth-Century Panama*. Westport, CT: Greenwood Publishing Group, 2001.

Taussig, Michael. *The Devil and Commodity Fetishism in South America*. Chapel Hill: The University of North Carolina Press, 1980.

Taussig, Michael. *My Cocaine Museum*. Chicago: University of Chicago Press, 2004.

Taylor, Peter. "World-System Theory." In *The Dictionary of Human Geography*, edited by R. J. Johnston, Derek Gregory, Geraldine Pratt, and Michael Watts, 901–903. Malden, MA: Blackwell Publishing, 2000.

Teclaff, Ludwik A. *The River Basin in History and Law*. The Hague: Martinus Nijhoff, 1967.

Thompson, E. P. "The Moral Economy of the English Crowd in the Eighteenth Century." *Past and Present* 50, no. 1 (1971): 76–136.

Tsing, Anna. *Friction: An Ethnography of Global Connection*. Princeton. NJ: Princeton University Press, 2005.

Vandergeest, Peter, and Nancy Lee Peluso. "Territorialization and State Power in Thailand." *Theory and Society* 24, no. 3 (1995): 385–426.

Vierba, Ezer. *The Committee's Report: Punishment, Power and Subject in 20th Century Panamá*. Dissertation, Department of History, Yale University, 2013.

von Schnitzler, Antina. "Citizenship Prepaid: Water, Calculability, and Techno-Politics in South Africa." *Journal of Southern African Studies* 34, no. 4 (2008): 899–917.

Wadsworth, Frank. "Deforestation: Death to the Panama Canal." In *US Strategy Conference on Tropical Deforestation*, 22–25. Washington, DC: US Department of State and US Agency for International Development, 1978.

Wali, Alaka. *Kilowatts and Crisis: Hydroelectric Power and Social Dislocation in Eastern Panama*. Boulder, CO: Westview Press, 1989.

Wallerstein, Immanuel. *The Modern World-System I: Capitalist Agriculture and the Origins of the European World-Economy in the Sixteenth Century*. New York: Academic Press, 1974.

Walley, Christine J. *Rough Waters: Nature and Development in an East African Marine Park*. Princeton, NJ: Princeton University Press, 2004.

Weil, Thomas E., Jan Knippers Black, Howard I. Blutstein, David S. McMorris, Frederick P. Munson, and Charles Townsend. *Area Handbook for Panama*. Washington, DC: Government Printing Office, 1972.

Weisman, Alan. *The World Without Us*. New York: Picador, 2007.

Whatmore, Sarah. *Hybrid Geographies: Natures, Cultures, Spaces*. London: SAGE, 2002.

White, Richard. *The Organic Machine: The Remaking of the Columbia River*. New York: Hill and Wang, 1995.

Wolff, Jane. "Redefining Landscape." In *The Tennessee Valley Authority: Design and Persuasion*, edited by Tim Culvahouse, 52–63. New York: Princeton Architectural Press, 2007.

Wood, Denis. *Rethinking the Power of Maps*. New York: Guilford, 2010.

Index

Classification
biological, 209–212
export bananas and, 147–148
imperial state and, 15, 228n34
"natives," as ambiguous category, 91,
145, 249n27
politics of land cover and, 30–32,
55–58
racial and national, of Panama Canal
labor, 110, 115–116, 177,
246–247n3
US overseas territories, 96
Climate
agriculture and, 31, 140, 148
forests influences on, 43, 49, 231n14
Panama Canal construction and, 86,
99, 100, 115, 177
race and, 87, 140 (see also
Determinism, environmental)
water supply for the Panama Canal
and, 41–42, 219–220
Coclé province (Panama), 150, 198
Colombia
concession to French canal company
in Panama, 85, 89
concession to Panama Railroad
Company, 173
Panamanian independence from, 90
political history with Panama, 77–78,
85, 89–90
US negotiations with for canal
concession, 89–90
watershed management in, 50–51,
235n46
Colombian Cultivators Law (1882),
112. See also Property
Colón
banana shipments from, 82, 84,
145–146, 149, 181
canal laborer settlement in, 109, 143
establishment of, 80
Free Trade Zone (Zona Libre), 16, 30,
124, 162

infrastructure built by US government
in, 108–109, 117–118
port of, 82, 84, 137, 145–146, 149
province of, 127, 200
roads from rural interior to, 17,
169–170, 172, 174–175
wage labor in, 30, 123–125, 181
Colonialism
anti-, 45, 77–78 (see also Nationalism,
Panamanian and the canal)
bananas and plantation economy
during, 135, 247
infrastructure and, 118, 179
interoceanic transportation and, 6,
74–77, 99
Spanish, 74–77, 91–92
US occupation of Panama as
"informal," 224n5
Colonization. See also Colonialism;
Imperialism
by campesinos, "spontaneous,"
21–22, 166, 181–183, 188–191,
200–203
programs of Panamanian state, 17,
187, 189–191, 195–198, 219
Columbian exchange. See Crosby,
Alfred; Ecological imperialism
Columbus, Christopher, 74–75
Competing infrastructures. See
Conflict, at the intersection of
infrastructures
Compradores (middlemen), banana
trade, 133, 147–148, 150–151, 155,
181
Concrete and modernity, 172–173, 179,
253n13. See also Mud; Poetics of
infrastructure
Conflict, at the intersection of
infrastructures, 6, 14–18, 21–23,
50–58, 166, 202–203
Connection (global) and infrastructure,
5, 6, 8, 11–23, 49, 92, 101, 118,
128, 130, 151–152, 162, 166, 168,

292

Index

Property (cont.)
issues in the Panamanian interior,
188–189 (see also Agrarian reform;
Tierra libre)
issues related to watershed
management, 44–47, 49, 55–56
Public health. See Panama Canal
Company Department of Sanitation
Pueblos perdidos (lost towns), 122–130,
146, 169

Race and racism
discourses of cultural "backwardness,"
34–35, 118, 135, 140, 142, 144, 154,
166, 171–172, 190, 200–201,
250n50
discourses of laziness and
irrationality, 87, 118, 140, 142, 144,
250n50
in the Panama Canal Zone, 6, 45, 94,
98, 110, 115–116, 118–119,
126–128, 133, 135, 140, 142–144,
146, 154, 177, 246–247n3 (see also
Gold and silver roll system)
Railroads
bananas and, in Central America,
134–136, 248n12
etymology of "infrastructure" and, 11
transcontinental in Panama
(see Panama Railroad)
transcontinental in United States,
82–83
standards and, 226n18 (see also
Gateways)
Rainfall. See Hydrology; Seasons (rainy
and dry)
Rainy season. See Seasons (rainy and
dry)
Rastrojo 22, 31–32, 35, 38–39, 55, 58.
230n3. See also Forest Law 13; Monte
Rationality. See Economic rationality;
Government rationality
Reclus, Armand, 85

Reforestation, 31, 47, 67, 145, 239n10,
250n42. See also Deforestation;
Rastrojo
Region-making, xii, 13, 22, 44, 48–49,
52, 54, 59–68, 107–108, 117, 132,
156, 179, 183, 200, 202, 203,
237n3. See also Maps and mapping;
Enrollment; Territorial politics
RENARE (Dirección Nacional de
Recursos Naturales Renovables). See
Environmental agencies (Panama)
Representation. See Discourse
Resistance, Panamanians and the canal,
7–8, 45–46, 90, 179
Revolution
green, agricultural development, 221
(see also Demanding environments)
in Guardia Nacional discourse,
199–200
Panamanian, national independence
from Colombia, 89–90
Rhizomes, 133–135, 148, 151–153, 208,
211, 216–217. See also Adventitious
growth; Bananas; Water hyacinth
River basin. See Watershed
River basin planning. See Watershed
management
Roads
administration (US), 176–179
agency (Panama), 170–171, 174, 179,
198–200, 202–203, 255n39
as always unfinished, 183, 187–188,
204, 219 (see also Infrastructure, as
processual)
Boquerón and, 163–165, 185–187,
191–195, 198, 255n1
Canal Zone, 134, 159–160, 175
colonial era (Camino de Cruces and
Camino Real), 75–77, 168, 175, 179,
191
development and, 13–14, 17–18,
22–23, 34, 78, 150, 169–182,
186–187, 195–203

Sedimentation
historical inertia of transportation
projects in Panama as, 18, 72–73,
91–92
of Panama Canal storage reservoirs
and channels, 4, 34, 40, 48, 56, 60,
118, 219
Shifting cultivation. See *Roza*
agriculture; Swidden agriculture
Ships and shipping. *See also* Global
infrastructure
colonial era, 74–77
containers, 12–14, 19–21, 30, 60,
72, 160, 162, 226n18, 226n20,
246n2
draft restrictions and, Panama Canal,
4, 40–41, 44, 231nn9–11
growth of maritime transportation
after Second World War, 41
Panama Canal as chokepoint for,
18–19, 163, 228n39
Panama Canal expansion project,
18–21
Panama Canal transit statistics and
key trade routes, 8
standards and Panama Canal locks,
18–21, 101–103, 226n18
steamships, 78–82, 84, 100, 146–152,
169, 210, 212
Shipping container, 11–14, 19–21, 30,
60, 72, 160, 162, 226n18, 226n20,
246n2. *See also* Gateways; Global
Infrastructure; Standards
Siltation. *See* Sedimentation
Slash-and-burn agriculture. See *Roza*
agriculture; Swidden agriculture
Slaves and slavery, 75–76, 87, 135,
247n8
Soberania National Park (Panama), 54
Society of the Chagres (Canal Zone),
114–115
Soo Locks, 100

Sovereignty, Panama and, 7–8, 16,
45–46, 77–78, 90, 94–95, 144,
171–173, 179, 190, 204, 224n5
Space. *See* Circulation; Global; Maps
and mapping; Place; Region-making;
Ships and shipping; Surveys;
Territorial politics; Uneven
development
Spain, 6, 74–77, 91–92
"Sponge effect," 42–43, 60, 231n14,
233n21. *See also* Forests,
hydrological effects of
"Spontaneous" campesino
colonization, discourse of, 21–22,
166, 181–183, 188–191, 200–203
Spooner Act of 1902 (US), 102
"Squatter problem," Canal Zone,
44–47, 111–115, 126, 202, 233n24
Standard Fruit, 147–148, 150
Standards
bananas and, 147–148, 155
global infrastructure and, 11–14,
18–21, 101, 226n18 (*see also*
Gateways)
Panamax ships and Panama
Canal locks, 18–21, 101–103,
226n18
road construction, 177, 182 (*see also*
Transístmica)
shipping containers, 12–14, 19–21,
30, 60, 72, 160, 162, 226n18,
226n20, 246n2
Star and Herald (newspaper), 103,
170
Star, Susan Leigh, 225n17, 237n10
State, the. *See also* Panama, Republic of;
Panama Canal Zone
as fragmented, 155–156, 174
global theory and, 8–11
known and assessed through
infrastructure, 15
materiality of, 7–8, 15, 227n30

rationalities of (*see* Government
   rationality)
role in development, 13–14
territorial politics of, 7, 15, 202
watershed management and, 48–51
State Department, US. *See* US
   Department of State
Steamships, 78–82, 84, 100, 146–152,
   169, 210, 212
Stevens, John, 96, 100
Streamflow, Panama Canal water
   management and, 4, 40–43, 60,
   65–66, 79, 87, 93, 103–105, 192,
   209. *See also* Forests, hydrological
   effects of; Panama Canal Company,
   Section of Meteorology and
   Hydrology
STS. *See* Science and technology
   studies
"*Sucia*" (dirty) landscapes, 128–129. *See*
   Maintenance; Weeds
Suez Canal, 18, 85–87, 100
Summit-level reservoir. *See* Alajuela
   Lake; Gatun Lake; Locks, Panama
   Canal
Surveys
   Biological Survey of the Panama
      Canal Zone, 209–211
   campesino migration surveys, 191
   Chagres River basin surveys, 75, 99,
      104–107, 243n32
   forest guard survey of Panama Canal
      watershed's human population,
      53–54
   French canal company survey of
      potential routes, 85
   Transístmica (transisthmian highway)
      survey, 175
Sustainable development, 30, 34–35,
   56–58
Swidden agriculture, 30–31, 55–56, 74,
   209–211. See also *Roza* agriculture

Taft, William, 110
Taylor, William A., 132, 137, 140–142,
   148
Technology. *See* Infrastructure; Design
Tennessee Valley Authority (TVA),
   50–52, 235n46. *See also* Watershed
   management
Territorial politics
   cities, ecology and, 227n29
   infrastructures and, 14–15
   maps and, 61
   Panama Canal Zone and, 7–8, 16, 22,
      45, 91, 94–96, 106–119, 132, 156
   Republic of Panama and, 7–8, 17–18,
      22–23, 34, 45–46, 168–175, 179,
      187, 189–191, 195–202, 219 (*see also*
      "Conquest of the Jungle"; Roads)
   theorizing the state and, 7–8, 15, 203,
      227n30
   US empire and, 15–16, 78, 88–90, 94,
      96
Third Locks project, 189
Tierra libre (free land). *See* Agrarian
   reform; Landlessness
Topography, 49, 73, 107. *See also*
   Geology of Panama
   in region of the Panama Canal, 54,
      73, 100, 140, 163, 178, 190
   territorial politics of the Canal Zone
      and, 107, 113–114, 124
   watersheds as "unity" of, 49, 54
Torrijos, Omar, 34, 45–46, 201
Torrijos-Carter Treaties (1977)
   environmental provisions and
      management issues, 46–48, 54, 63,
      66–68
   signing and 1979–1999 transition
      period, 16, 46, 54, 234n30
Tourism of the Panama Canal, 2,
   97–99, 160
Transisthmian highway. *See*
   Transístmica

298 Index

West Indians
agriculture in Canal Zone, 140–156
banana cultivation, history of,
135–136, 145, 247n8, 247n10,
249n33
employers' preference for over
Spanish-speaking "natives," 87,
141
labor migration across Central
America after opening of canal, 110,
143
labor migration to Panama during
nineteenth century, 87, 90–91
recruitment as labor, 80, 87, 123, 136,
143, 247n10
labor strikes by, 143, 249n35
laziness and irrationality among,
racialized discourse of, 118, 140,
142, 144, 250n50
low status as "silver roll" laborers in
US Panama Canal project, 98, 138,
143
resettlement from rural Canal Zone,
45, 110–116, 118–119, 249n29
unemployment, after opening of
Panama Canal, 110–111, 132–134,
142–143
Workman, newspaper in Panama for,
111, 172
West Indies. *See* Caribbean; West
Indians
Wet season. *See* Seasons (rainy and dry)
Wetmore, Alexander, 185–186
Wilderness, appearance masking
human history, 28, 61, 67, 191–192,
239n9
Wolff, Jane, 51–52
Work, infrastructural, 5, 11, 21, 57, 72,
92, 107–108, 219–221. *See also*
Infrastructure, as processual;
Maintenance
Workman (newspaper), 111, 172
World Bank, 13

World-system theory, 9–10, 92, 224n10
World War I, 142, 199
World War II
colonization of Panama's forested
frontier after, 187
expansion of maritime transportation
after, 41
global dissemination of Tennessee
Valley Authority model after,
235n45
infrastructure concept, ascendance
after, 226n24
Panama Canal defense projects, 123,
168, 175–177, 189 (*see also*
Transístmica)
as transitional moment for Gatun
Lake communities, 123, 128
watershed management and
development after, 50, 235n45

Yellow fever, 88, 116, 136. *See also*
Panama Canal Company,
Department of Sanitation

Zona Libre. *See* Free Trade Zone; Colón
Zone. *See* Panama Canal Zone

Printed in the United States
by Baker & Taylor Publisher Services